Smart Innovation, Systems and Technologies

Volume 74

About this Series

The Smart Innovation, Systems and Technologies book series encompasses the topics of knowledge, intelligence, innovation and sustainability. The aim of the series is to make available a platform for the publication of books on all aspects of single and multi-disciplinary research on these themes in order to make the latest results available in a readily-accessible form. Volumes on interdisciplinary research combining two or more of these areas is particularly sought.

The series covers systems and paradigms that employ knowledge and intelligence in a broad sense. Its scope is systems having embedded knowledge and intelligence, which may be applied to the solution of world problems in industry, the environment and the community. It also focusses on the knowledge-transfer methodologies and innovation strategies employed to make this happen effectively. The combination of intelligent systems tools and a broad range of applications introduces a need for a synergy of disciplines from science, technology, business and the humanities. The series will include conference proceedings, edited collections, monographs, handbooks, reference books, and other relevant types of book in areas of science and technology where smart systems and technologies can offer innovative solutions.

High quality content is an essential feature for all book proposals accepted for the series. It is expected that editors of all accepted volumes will ensure that contributions are subjected to an appropriate level of reviewing process and adhere to KES quality principles.

More information about this series at http://www.springer.com/series/8767

Gordan Jezic · Mario Kusek
Yun-Heh Jessica Chen-Burger
Robert J. Howlett · Lakhmi C. Jain
Editors

Agent and Multi-Agent Systems: Technology and Applications

11th KES International Conference, KES-AMSTA 2017
Vilamoura, Algarve, Portugal, June 2017
Proceedings

Springer

Editors
Gordan Jezic
Faculty of Electrical Engineering and Computing
University of Zagreb
Zagreb
Croatia

Mario Kusek
Faculty of Electrical Engineering and Computing
University of Zagreb
Zagreb
Croatia

Yun-Heh Jessica Chen-Burger
School of Mathematical and Computer Sciences
Heriot-Watt University
Edinburgh
UK

Robert J. Howlett
Fern Barrow
Bournemouth University
Poole, Dorset
UK

Lakhmi C. Jain
University of Canberra
Canberra, ACT
Australia

ISSN 2190-3018 ISSN 2190-3026 (electronic)
Smart Innovation, Systems and Technologies
ISBN 978-3-319-86615-4 ISBN 978-3-319-59394-4 (eBook)
DOI 10.1007/978-3-319-59394-4

Printed on acid-free paper

This Springer imprint is published by Springer Nature
The registered company is Springer International Publishing AG
The registered company address is: Gewerbestrasse 11, 6330 Cham, Switzerland

Preface

This volume contains the proceedings of the 11th KES Conference on Agent and Multi-Agent Systems—Technologies and Applications (KES-AMSTA 2017) held in Vilamoura, Algarve, Portugal, between June 21 and 23, 2017. The conference was organized by KES International, its focus group on agent and multi-agent systems and University of Zagreb, Faculty of Electrical Engineering and Computing. The KES-AMSTA conference is a subseries of the KES conference series.

Following the successes of previous KES Conferences on Agent and Multi-Agent Systems—Technologies and Applications, held in Puerto de la Cruz, Tenerife, Spain (KES-AMSTA 2016), Sorrento, Italy (KES-AMSTA 2015), Chania, Greece (KES-AMSTA 2014), Hue, Vietnam (KES-AMSTA 2013), Dubrovnik, Croatia (KES-AMSTA 2012), Manchester, UK (KES-AMSTA 2011), Gdynia, Poland (KES-AMSTA 2010), Uppsala, Sweden (KES-AMSTA 2009), Incheon, Korea (KES-AMSTA 2008), and Wroclaw, Poland (KES-AMSTA 2007), the conference featured the usual keynote talks, oral presentations, and invited sessions closely aligned to the established themes of the conference.

KES-AMSTA is an international scientific conference for discussing and publishing innovative research in the field of agent and multi-agent systems and technologies applicable in the digital and knowledge economy. The aim of the conference was to provide an internationally respected forum for both the research and industrial communities on their latest work on innovative technologies and applications that is potentially disruptive to industries. Current topics of research in the field include technologies in the area of mobile and cloud computing, big data analysis, business intelligence, artificial intelligence, social systems, computer embedded systems, and nature-inspired manufacturing. Special attention is paid on the feature topics: business process management, agent-based modeling and simulation, and anthropic-oriented computing.

The conference attracted a substantial number of researchers and practitioners from all over the world who submitted their papers for main track covering the methodologies of agent and multi-agent systems applicable in the digital and knowledge economy, and three invited sessions on specific topics within the field.

Submissions came from 12 countries. Each paper was peer-reviewed by at least two members of the International Programme Committee and International Reviewer Board. Twenty-three papers were selected for oral presentation and publication in the volume of the KES-AMSTA 2017 proceedings.

The Programme Committee defined the main track entitled Agent and Multi-Agent Systems and the following invited sessions: Agent-based Modeling and Simulation (ABMS), Business Process Management (BPM), and Anthropic-Oriented Computing (AOC).

Accepted and presented papers highlight new trends and challenges in agent and multi-agent research. We hope that these results will be of value to the research community working in the fields of artificial intelligence, collective computational intelligence, robotics, dialogue systems, and, in particular, agent and multi-agent systems, technologies, tools, and applications.

The Chairs' special thanks go to the following special session organizers: Dr. Roman Šperka, Silesian University in Opava, Czech Republic, Prof. Salvatore Distefano, University of Messina, Italy, and Kazan Federal University, Russia, Max Talanov, Kazan Federal University and Innopolis University, Russia, Prof. Jordi Vallverdú, Universitat Autònoma de Barcelona, Spain, and Evgeni Magid, Kazan Federal University, Russia, for their excellent work.

Thanks are due to the programme co-chairs, all programme and reviewer committee members, and all the additional reviewers for their valuable efforts in the review process, which helped us to guarantee the highest quality of selected papers for the conference.

We cordially thank all authors for their valuable contributions and all of the other participants in this conference. The conference would not be possible without their support.

April 2017

Gordan Jezic
Mario Kusek
Yun-Heh Jessica Chen-Burger
Robert J. Howlett
Lakhmi C. Jain

KES-AMSTA 2017 Conference Organization

KES-AMSTA 2017 was organized by KES International—Innovation in Knowledge-Based and Intelligent Engineering Systems.

Honorary Chairs

I. Lovrek	University of Zagreb, Croatia
L.C. Jain	University of South Australia, Adelaide

Conference Co-chairs

G. Jezic	University of Zagreb, Croatia
J. Chen-Burger	Heriot-Watt University, Scotland, UK

Executive Chair

R.J. Howlett	Bournemouth University, UK

Programme Co-chairs

M. Kusek	University of Zagreb, Croatia
R. Sperka	Silesian University in Opava, Czech Republic

Publicity Chair

P. Skocir	University of Zagreb, Croatia

International Program Committee

Koichi Asakura	Daido University, Japan
Marina Bagić Babac	University of Zagreb, Croatia
Costin Badica	University of Craiova, Romania
Dariusz Barbucha	Gdynia Maritime University, Poland
Iva Bojic	MIT, USA
Zoran Budimac	University of Novi Sad, Serbia
Frantisek Capkovic	Institute of Informatics, Slovak Academy of Sciences, Slovakia
Yun-Heh (Jessica) Chen-Burger	The Heriot-Watt University, Edinburgh, UK
Angela Consoli	Defence Science and Technology Group, Australia
Matteo Cristani	University of Verona, Italy
Ireneusz Czarnowski	Gdynia Maritime University, Poland
Salvatore Distefano	University of Messina, Italy
Nicola Dragoni	Technical University of Denmark, Denmark and Örebro University, Sweden
María del Rosario Baltazar Flores	Instituto Tecnológico de León, México
Arnulfo Alanis Garza	Instituto Tecnológico de Tijuana. México
Natalya Garanina	Institute of Informatics Systems, Russia
Paulina Golinska-Dawson	Poznan University of Technology, Poland
Anne Håkansson	Software and Computer Systems ICT Information and Communication Technology, KTH Royal Institute of Technology, Kista, Sweden
Chihab Hanachi	University of Toulouse 1 Capitole - IRIT Laboratory, France
Tzung-pei Hong	National University of Kaohsiung, Taiwan
Mirjana Ivanovic	University of Novi Sad
Dragan Jevtic	University of Zagreb, Croatia
Vicente Julian	Universitat Politecnica de Valencia, Spain
Radosław Piotr Katarzyniak	Wroclaw University of Technology, Poland
Arkadiusz Kawa	Poznan University of Economics and Business, Poland
Petros Kefalas	The University of Sheffield International Faculty, City College
Adrianna Kozierkiewicz-Hetmańska	Wroclaw University of Science and Technology, Poland
Konrad Kułakowski	AGH University of Science and Technology, Poland
Setsuya Kurahashi	University of Tsukuba, Japan
Kazuhiro Kuwabara	Ritsumeikan University, Japan

Mario Kusek	University of Zagreb, Croatia
Lenin G. Lemus-Zúñiga	Universitat Politècnica de València, España
Marin Lujak	University Rey Juan Carlos
Evgeni Magid	Kazan Federal University, Russia
Manuel Mazzara	Innopolis University, Russia
Daniel Moldt	University of Hamburg, Germany
Cezary Orłowski	Gdansk School of Banking, Poland
Radu-Emil Precup	Politehnica University of Timisoara, Romania
Vedran Podobnik	University of Zagreb, Croatia
Rajesh Reghunadhan	Central University of Kerala, India
Ewa Ratajczak-Ropel	Gdynia Maritime University, Poland
Silvia Rossi	University of Naples Federico II, Italy
Nikolay Shilov	Innopolis University, Russia
Roman Šperka	Silesian University in Opava, Czech Republic
Darko Stipanicev	University of Split, Split, Croatia
Max Talanov	Kazan Federal University, Russia
Wojciech Thomas	Department of Software Engineering, Wroclaw University of Science and Technology (Politechnika Wrocławska), Wrocław, Poland
Krunoslav Tržec	Ericsson Nikola Tesla, Croatia
Taketoshi Ushiama	Kyushu University, Japan
Jordi Vallverdu	Universitat Autonoma de Barcelona, Spain
Bay Vo	Ho Chi Minh City University of Technology, Vietnam
Toyohide Watanabe	watanabe@nagoya-u.jp
Izabela Wierzbowska	Gdynia Maritime University, Poland
Mahdi Zargayouna	University of Paris-Est, IFSTTAR, France
Arkady Zaslavsky	CSIRO ICT Centre, Australia

Invited Session Chairs

Business Process Management

Roman Šperka — Silesian University in Opava, Czech Republic

Agent-Based Modelling and Simulation

Roman Šperka — Silesian University in Opava, Czech Republic

Anthropic-Oriented Computing

Salvatore Distefano	University of Messina, Italy, Kazan Federal University, Russia
Jordi Vallverdu	Universitat Autonoma de Barcelona, Spain
Evgeni Magid	Kazan Federal University, Russia
Max Talanov	Kazan Federal University, Russia

Contents

Business Process Management

Agent and Multi-Agent Systems

Personalized HealthCare and Agent Technologies

Mirjana Ivanović[1(✉)] and Srđan Ninković[2]

[1] Department of Mathematics and Informatics, Faculty of Sciences, University of Novi Sad, Novi Sad, Serbia
mira@dmi.uns.ac.rs
[2] Department of Orthopedic Surgery and Traumatology, Clinical Center – Vojvodina, University of Novi Sad, Novi Sad, Serbia
srdjan.ninkovic@yahoo.com

Abstract. Remarkable gains in life expectancy and declines in fertility have led current society to an ageing global population. Different stakeholders, researcher communities and policy makers invest serious efforts to develop intelligent and smart environments that have to support as much as possible independent living of old population. As necessary prerequisite for these efforts rapid and fascinating development in ICT offers wide range of new technologies including wearable, 3D sensors and smart environments. These new technologies provide rich complex data from living environment and give the opportunity to learn and analyze them in order to discover the patient's preferences, traits, and states. Further research efforts are oriented to personalized healthcare and development of sophisticated e-coaching facilities to obtain proper recommendations and advices to patients in order to increase their wellbeing.

Among different artificial intelligence methods and techniques agent technologies significantly influence and support different medical domains. The use of agents and multi agent systems in healthcare has also opened the ways to find out new applications like personalized and socialized healthcare platforms and systems with tailored recommendation capabilities. In this paper opportunities and challenges that agent technologies offer in personalized healthcare are discussed and presented.

1 Introduction

Remarkable gains in life expectancy and declines in fertility have led current society to an ageing global population. Life expectancy has increased to 70 years or more in many countries. The ageing of populations has led to changes in the prevalence of disease and disability.

Consequently in last decade there is growing need to supply constant healthcare monitoring and adequate support of elderly and disabled people.

Different platforms and tools for monitoring, smart support and making personalized recommendations to old people and patients have been developing. One of emerging research and application area is Ambient Intelligence (AmI). Ambient

G. Jezic et al. (eds.), *Agent and Multi-Agent Systems: Technology and Applications*, Smart Innovation, Systems and Technologies 74,
DOI 10.1007/978-3-319-59394-4_1

Intelligence for healthcare monitoring and personalized support is a promising research direction to provide efficient medical services for old and disabled patients.

New ICT technologies including wearable technologies (health watches and smartphones), 3D sensors and smart environments provide rich complex data and give the opportunity to learn and analyze them in order to discover the patient's preferences, traits, states and context.

Concepts and techniques of artificial intelligence (AI), data mining, and agent technology are unavoidable in AmI applications and platforms where sensors technology to monitor patient in the home and community settings are used.

The main objective of AmI is to try to adapt the technology to the patient's needs by different omnipresent devices which mutually communicate in a ubiquitous way. Various architectures and platforms of AmI application for health and wellness, home rehabilitation, assessment of treatment have been developing (Salih and Abraham 2013).

During usual patient activities wide range of data has been collected and later it must be processed, analyzed and later intelligent reasoning and proper decision making can help patients in everyday activities.

Software (and intelligent) agents represent one of widely applicable technology in different areas and domains. Agent technologies provide the right architecture for two major computing areas widely used in medical systems and environments: artificial intelligence (AI) and pervasive (seamless) computing.

In literature, the use of intelligent software agents has been proposed to deal with a variety of medical and health related problems: patient and treatment information access, community care, decision support systems (DSS), patient scheduling, training, hospital management, elder citizen care, self-care and automatic health monitoring (Iqbal 2016).

In this paper a quick overview of several contemporary trends in development of medical and healthcare systems will be presented. The rest of the paper is organized as follows. Section 2 is devoted to basic characteristics of personalization in medicine and healthcare. E-coaching in medical domains is presented in Sect. 3. Section 4 discusses some possible roles of agents in personalized medicine and e-coaching. Last section brings some concluding remarks.

2 Personalization in Medicine and HealthCare

Rapid increase of use smartphones and 3G and 4G networks has triggered expanded use of health devices and influenced a lot of different medical aspects like healthcare of aging people. Low-cost sensors have led to their integration into a wide range of wearable devices. So through smartphones and tablets patients can access different health data and monitor their daily activities. In fact the patients are moving into different technological interconnected worlds.

Healthcare wearables are those wearables for measuring metrics that are assumed to provide an indication of a patient's health and state of wellbeing. Recently there has been the emergence of wearables that are able to monitor detailed clinical metrics, such as blood pressure, heart function, glucose and insulin levels, and medicine intake and so on.

Additional advantage of use of ICT devices in everyday life is that patients can share their results and behavior through social networks and engage in lifestyle improvement games with their peers. Such hyper-connected patients open the new opportunities and directions of research that can help ageing population to cope with everyday activities smoothly and independently: ubiquitous and smart environments, personalized medicine, healthcare e-coaching.

In fact a lot of platforms developed to boost patients in their leaving space function as AmI environments. AmI puts together several crucial resources to provide flexible and intelligent services to patients acting in their environments: Pervasive Ubiquitous Computing and Artificial Intelligence (AI), Networks, Sensors and Unobtrusive Human Computer Interfaces (HCI). Complex context data can be collected by distributed sensors throughout the environment. When analyzing sensor data two strategies can be recognized: distributed or centralized. In the distributed model, each sensor has onboard processing capabilities and performs local computation before communicating partial results to other nodes in the sensor network (Salih and Abraham 2013). In the centralized model data is transmitted to a central server, which fuses and analyzes the received data.

Nevertheless which approach is adopted in analyzing sensors data it is possible to create a computational challenge for modeling several AI and data mining methods. But there exist another slightly different research direction that is based on and incorporate agent technologies in creating platforms for healthcare support.

Personalization in different e-services and systems is trend that exists for more than a decade. Personalization consists of tailoring a service or a product to accommodate specific individuals or groups. Personalization in different medical systems is newest trend. Personalized medicine and healthcare promise prediction, prevention and treatment of illness that is targeted to patients' needs. Personalized medicine is oriented towards the collecting of information from the patient in order to better tailor his/her needs. So to raise quality of life especially for elderly people it is necessary to orient research and activities towards more targeted prediction, prevention and treatment of illness. New technologies for detailed biological profiling of patients at the molecular level have been crucial in initiating the move to personalized medicine. To obtain more reliable mechanisms and support for patients it is necessary to continuously develop new technologies for collecting and properly analyzing complex medical, personal, environmental and behavioral data (Harvey et al. 2012).

There is also demand for qualitative increase: new types of data such as data on the patient's everyday leaving environment (nutrients, the microbiome, toxin exposure) are being seen as important for understanding biological functioning. More information is required in both the research setting, and in healthcare practice and also development of tools to monitor and manage patient's health status as part of their everyday life as a form of self-monitoring. Novel technologies and devices are being employed to facilitate health monitoring and they are becoming more sophisticated, capable of tracking several physiological variables and communicating them to a mobile smartphone or other computerized device. A step further represent the new technologies and gadgets that will allow the ongoing monitoring of functional status in real time, allowing fine tuning of therapy or adjustment of lifestyle to achieve the patient's health goals. Within self-monitoring and collected data from different devices it is necessary to obtain efficient ways of analyses of the data to identify areas for improvement, provide education on how to achieve desired health goals, and gamification to increase engagement, as well as encouraging individuals to share their achievements with friends, compete and collaborate; providing further motivation to continue improving (usually via social network sites).

Presence of the diverse needs of an ageing society with use of wide range of ICT platforms, tools, and intelligent data mining algorithms is a key challenge in achieving following demands (PwC Global 2015):

1. Help older people to stay independent and healthy for as long as possible
2. Help older people to manage simple chronic conditions
3. Help older people that complex co-morbidities remain independent
4. Help older people to minimize the time they have to spend in hospital.

Technology, including telehealth, wearable devices, and sensor driven detection software in homes, are increasingly helping older people and their relatives to engage and communicate with service providers on their own terms. New technology, including different smart devices and wearables connected to the internet, collects more and more data (big and complex data), increasingly outside existing care providers. As such, the individual becomes the central node in the use of his or her own data, and therefore involved in the seamless delivery of the personalized services they need.

So personalized approach to care of elderly patients has to include different interconnected components, which will be incorporated in unique platform and obtain efficient and prompt support:

- Self-rated quality of life
- Family, community & peer support
- Smart homes
- Health apps and remote wellbeing monitoring.

Agent technologies play important role and can significantly help in developing higher-quality services.

3 E-coaching in Medical Domains

Coaching is new trend in different areas and aspects of human everyday activities. It generally can promote relationships, feedback, care, conversation, collaboration, answers, and bonding between different persons, groups, and communities. Great coaches are motivators, tending to boost individual's confidence and other emotional and behavioral issues. Extensive use of the internet advanced different aspects of coaching and introduced so called e-coaching. E-coaching performs the process online and greatly expands the possibilities. So online experiences are the essential way of supporting the coaching relationships (Rossett and Marino 2005). One important component of e-coaching is reducing costs, while providing encouragement, information, and connection to networks of people and content.

Experts from different disciplines and areas agree that there is a strong link between behavior and health. Healthy lifestyle can prevent many diseases. Personalizes e-coaching can play significant role in supporting people to achieve their health goals and properly maintain their healthy behavior. This shift towards more personalized healthcare is reflected in the change of focus from a disease-centered approach towards a patient-centered approach. This obtains empowering patients to take an active role in the decisions about their own health. There are different sources of data for a patient starting from rather traditional ones like electronic healthcare records towards completely new types and forms of data like as data obtained from the patient's environment (nutrients, the microbiome, toxin exposure, gait and more others). Artificial intelligent techniques, effective data mining algorithms can help to separate relevant from irrelevant information and discover significant cognitive, emotional and behavioral patterns. From this point of view (Rutjes et al. 2016) e-coaching includes many aspects, e.g. persuasion, behavior change, personal contact and a type of recommendations.

Personalization, contextualization and frequent adaptation are necessary prerequisites of the e-coaching process. According to that seems that agent technologies are perfect candidates to take a role of an e-coaches that supports automated self-help therapies (Beun et al. 2016).

Important aspect of e-coaching in the area of personalized medicine is oriented to emotional and cognitive technologies and includes following essential components:

- Wearables
- Cognitive Health (that are based on Cognitive Enhancement and Cognitive Assistance)
- Remote Patient Monitoring (for Activity Detection)
- Medication adherence (that incorporates: Different Devices, Reminder Systems, Coaching and Advising, Coordination Systems)

4 Agents and Their Role in Personalized and E-coaching Medicine

The most popular application of wearable technology recently has been in the area of health and wellness. Different wearable devices collect a pile of data about patient and his/her environment. All the data from the different devices set in the environment are collected but it is necessary to apply AI and Machine Learning (ML) techniques to analyze this data comprehensively. Also the patient is able to view the instant and historical data on their mobile devices. Such platforms and systems also have to provide the real-time hands-free feedback and instructions through the sophisticated user interfaces (visually, acoustically and tactilely). Sophisticated user interfaces that facilitate HCI more and more are realized as different personal virtual and visual agents (avatars).

The development of AmI-based software requires creating increasingly complex and flexible applications. Autonomous decision making agents that incorporate learning mechanisms, and are able to respond to events by (pre)planning in execution time are excellent mechanisms to be incorporated and support activities in AmI healthcare environments. With good reasoning and planning mechanisms agents facilitate acquiring data from different devices that patients use but also the data from their everyday living environment. They are also good mechanisms that support straight coordination and communication among wireless medical devices.

The successful use of intelligent agents in healthcare has attracted researchers to apply this emerging software engineering paradigm in more advanced and complex applications. The multi agent systems have been applied from single healthcare activity like knowledge-based medical system to complex, multi-component based systems. The use of multi agent systems in healthcare domain has also opened the ways to find out new applications like personalized and socialized healthcare systems with tailored recommendation capabilities.

Agents and Multi-Agent systems are good entities and concepts in abstraction tools, but predominantly for modeling devices and their interactions, to serve as personal assistants and recently as virtual e-coaches and advisors in patients' emotional and cognitive activities.

Full potential of AmI cannot be realized without sophisticated knowledge representation, reasoning, and AI and agent-oriented technologies.

In last two decades development of new theories, methods and technologies is emerging in order to support adaptive and personalized dialogues between a human and a (intelligent) software agent. Software agents can be incorporated in different health platforms and applications: supporting an expert's decision making (based on big and complex data collected from wearable devices and environmental sensors), accessing and making use of distributed data sources or the coordination of the execution of assistive technology for healthcare activities. Assistive technology can be supportive, preventive or responsive (Baskar 2014).

Apart from different practical challenges and possibilities software agents are essential facilitators in handling everyday queries about health that are perceived as meaningful and useful to the patients. The agent needs to have wide range of data and knowledge about the patient, the particular topic of the dialogue, and also necessary data about the physical and social environment in which patient is living. Moreover, the agent has to know how to be cooperative and be able to recognize patient's emotions and cognitive status and behave and express with empathy while conducting a dialogue activity. In some situations, it needs to approach to the decisions together with the patient and give him/her adequate recommendations and advices. The dialogue activities must be based on straightforward argumentation schemes and trust. The agent can adapt its moves to the patient's trail of reasoning, goals, and behave in an empathic way and adapt to the human's emotional state.

General structure of cognitive agent architecture (Baskar 2014) that could be used in personalized medical e-coaching as well is presented in Fig. 1.

The knowledge about the patient's activities involves observation of both personal and environment activity, obtained by the seamlessly integrated sensors as a part of a ubiquitous computing environment in which patient lives.

Different kinds of dialogues and e-coaching between virtual agent (e-coach) and patient could be performed, but some highly important in personalized medicine could be (Baskar 2014):

1. **Information-seeking dialogues**, where patient seeks the answer to some medical and health questions from virtual e-coach realized as intelligent software agent.
2. **Inquiry dialogues** when the patient collaborates with virtual e-coach to obtain an answer to specific question and to validate a claim about particular topic.

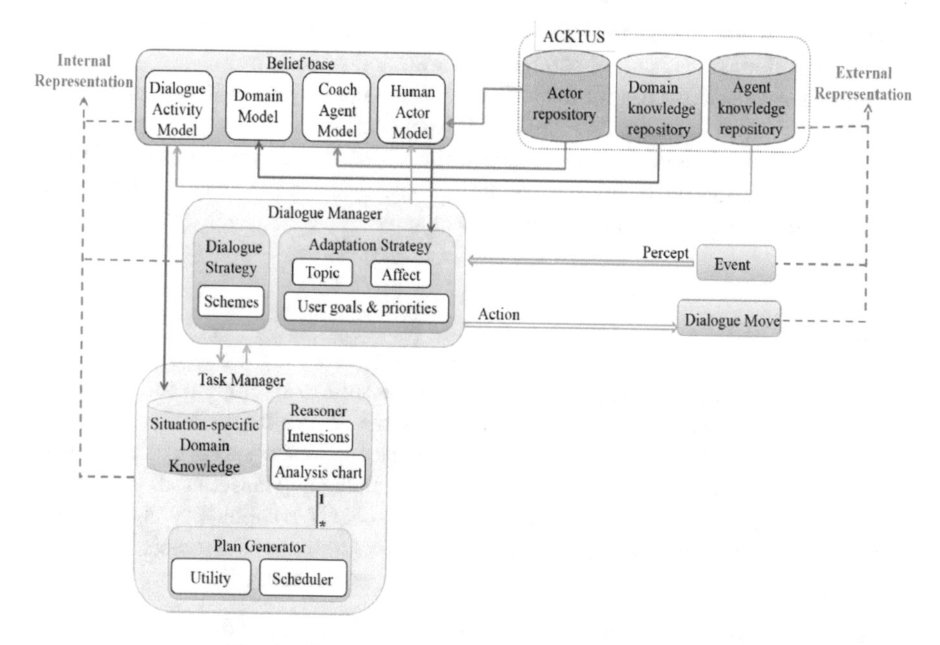

Fig. 1. Cognitive agent architecture (Baskar 2014)

3. **Deliberation dialogues**, in this dialogs both participants collaborate to decide what course of action should be adopted. For example in the healthcare, it could be the decision about interventions aimed at improving a patient's daily medical conditions.
4. **Persuasion dialogues** involve virtual e-coach seeking to guide and persuade patient to perform some activities and solve some situations adequately in accordance with his health state.

Software agents also can be used in other medical domains and to support different tools and platforms in patient-centered environments. Different sub-domains include ambient medical intelligence, medical data harvesting and collection, medical data management, knowledgebase and decision support systems, operational systems for healthcare, healthcare resource planning and management and so on (Iqbal 2016). A very brief description of these additional possibilities of use agents in medical systems is given below:

1. **Planning and resource management** - These systems focus on planning medical processes, monitoring of staff and performance measurement, patient health monitoring, hospital and clinical resources management.
2. **Decision support systems/knowledge base systems** - Such systems utilize knowledgebase and apply some type of data analysis techniques (using AI or machine learning), pattern recognition algorithms, and also might often use knowledge inference techniques.

3. **Data management systems** - These systems focus mainly on health data extraction, representation, organization, storage, retrieval, and presentation.
4. **Remote care/self care systems** - This area includes systems designed for automated patient monitoring remotely, and patient self-care.
5. **Multifunction systems** – There exists various systems that perform multiple tasks related to a complete healthcare solution. These systems are usually complex and may be composed of two or more sub-domains.

5 Conclusion

Changes in the healthcare sector have to address the shift to an empowered patient and the potential of wearable sensing data to personalize health treatment. Future of wearable application lies in the support of the user to make sense of data and get contextualized personalized feedback for behavior changes (Hänsel et al. 2015).

Emergent and rapid development of wide range of ICT components has significant influence on personalized medicine and virtual e-coaching. It also directs research in the area in the following directions (Gemert-Pijnen et al. 2016):

1. **Health analytics** - advanced methods (machine learning) and models to analyze Big and Complex Data.
2. **Predictive modeling** - to set up smart models to predict behaviors, to prevent diseases and to personalize healthcare.
3. **Visualization of data** – different tools support and facilitate presentation of data in meaningful way to support reliable decision making.
4. **Integration of mobile (hardware and software) technologies** – integration of different devices with data-platforms is important to enable automated services and to tailor feedback and recommendations.
5. **Personal communication and recommendations between patient and virtual e-coach** - In these area agent technologies definitely could play extremely important role. Virtual Human Agents are suitable components (have been developed on the bases of knowledge representation, cognitive and emotional modeling, natural language processing) to support and empower communication, personalization and increase motivation of patients.

References

Baskar, J.: Adaptive human-agent dialogues for reasoning about health. Licentiate Thesis, Umea University (2014)

Beun, R.J., Brinkman, W.-P., Fitrianie, S., Griffioen-Both, F., Horsch, C., Lancee, J., Spruit, S.: Improving adherence in automated e-coaching. In: Meschtscherjakov, A., De Ruyter, B., Fuchsberger, V., Murer, M., Tscheligi, M. (eds.) PERSUASIVE 2016. LNCS, vol. 9638, pp. 276–287. Springer, Cham (2016). doi:10.1007/978-3-319-31510-2_24

Gemert-Pijnen, J., Sieverink, F., Siemons, L., Braakman-Jansen, L.: Big data for personalized and persuasive coaching via self-monitoring technology. In: eTelemed 2016, Venice, Italy, 24–28 April 2016

Hänsel, K., Wilde, N., Haddadi, H., Alomainy, A.: Challenges with current wearable technology in monitoring health data and providing positive behavioural support. In: MobiHealth 2015 Proceedings of the 5th EAI International Conference on Wireless Mobile Communication and Healthcare, London, Great Britain, 14–16 October 2015

Harvey, A., Brand, A., Holgate, S., Kristiansen, L., Lehrach, H., Palotie, A., Prainsack, B.: The future of technologies for personalised medicine. New Biotechnol. **29**(6), 625–633 (2012). doi:10.1016/j.nbt.2012.03.009

Iqbal, S., Altaf, W., Aslam, M., Mahmood, W., Khan, M.: Application of intelligent agents in health-care: review. Artif. Intell. Rev. **46**, 83 (2016). doi:10.1007/s10462-016-9457-y

PwC Global, Connected and coordinated: Personalised service delivery for the elderly (2015). http://www.pwc.com/global-health. Accessed 1 Mar 2017

Rossett, A., Marino, G.: If coaching is good, then e-coaching is.... Training Dev. **59**(11), 46–49 (2005)

Rutjes, H., Willemsen, M., IJsselsteijn, W.: Understanding effective coaching on healthy lifestyle by combining theory- and data-driven approaches. In: Workshop on Personalization in Persuasive Technology at Persuasive (2016)

Salih, A., Abraham, A.: A review of ambient intelligence assisted healthcare monitoring. Int. J. Comput. Inf. Syst. Ind. Manag. Appl. **5**, 741–750 (2013)

Multiagent Environments for Dynamic Transportation Applications

Mahdi Zargayouna(✉)

Université Paris-Est, IFSTTAR, GRETTIA, Boulevard Newton, Champs sur Marne, 77447 Marne la Vallée Cedex 2, France
hamza-mahdi.zargayouna@ifsttar.fr

Abstract. Dynamic transportation applications have long been a domain of choice for the multiagent paradigm. Indeed, the presence of distributed entities, the highly dynamic character of these applications and the often presence of human actors in the system makes it very suitable for a multiagent design. This paper advocates for the primary consideration of multiagent environment design when dealing with such dynamic transportation applications. Transportation applications can greatly benefit from the use of the multiagent environment since most of them consider a dynamic geographical positioning of the system components. Indeed, the simultaneous consideration of the time and space dimensions makes the environment, which is shared and accessed by all the agents of the system, a candidate of choice to capture the dynamics of the application. The environment design can be envisioned at several levels of the system construction. It can be used as a medium for interaction between distributed entities. It could be used as a coordination entity of the system components. It can finally be designed as a mental model for the agents that they can use in their reasoning. We illustrate the possible uses of the environment with two transportation applications dealing with traveler information.

1 Introduction

The multiagent paradigm is proven to be a powerful model to design and implement transportation applications. Indeed, the multiagent approach deals with systems consisting of many physically or logically distributed interacting components that possess some level of autonomy. These components are able to perceive their environment and also react to changes in that environment in accordance to their goals. That is why the multiagent approach is adapted to the transportation domain since it facilitates an approach by analogy in a domain where the objective is the management of distributed entities. The authors in [1] list several reasons for the privileged use of multiagent systems in these applications, such as the natural and intuitive problem solving, the ability of autonomous agents for the modeling of heterogeneous systems, the ability to capture complex constraints connecting all problem-solving phases, etc. Indeed, the concept

G. Jezic et al. (eds.), *Agent and Multi-Agent Systems: Technology and Applications*, Smart Innovation, Systems and Technologies 74,
DOI 10.1007/978-3-319-59394-4_2

of an agent is well suited for the representation of travelers in transit or road traffic scenarios [2,3]. They are autonomous entities which are situated in an environment, adapt their behaviors to the dynamics they perceive and interact with others agents in order to achieve specific goals. For Parunak [4], "Agent-based modelling is most appropriate for domains characterized by a high degree of localization and distribution", which is the case for complex and dynamic transportation applications.

In the multiagent community, there is a growing conscience that the multiagent environment should be considered as a primary design abstraction, of equal importance as the agents. Models and architectures have been proposed in the literature for multiagent environments design, validated in a variety of application domains [5]. We believe that one of the domains of choice for the multiagent environment modeling is the transportation domain. Indeed, transportation applications always have some kind of representation for the environment, typically the transportation networks. The environment in transportation application has its own dynamics (e.g. traffic conditions, dynamic rules, weather, etc.), which advocates for its independent and explicit representation. Transportation systems are open, with entities joining and leaving the system (e.g. travelers, drivers, vehicles, regulators, etc.), generally in a nondeterministic way. The multiagent environment can also be the privileged interlocutor of the newcomer entities.

In this paper, we illustrate different design angles of the multiagent environment when dealing with transportation applications. The environment design can be envisioned at several levels of the system construction. It can be used as a medium for interaction between distributed entities. It could be used as a coordination entity of the system components. It can also be designed as a mental model for the agents that they can use in their reasoning. To illustrate the possible uses of the environment, two applications are considered: traveler information and information dissemination in disturbed transit networks.

The remainder of this paper is structured as follows. In Sect. 2, we present a generic design of multiagent environment, in the form of a specification language and the traveler information application built with the language, and using the environment to support agents interaction. In Sect. 3, we present a representation of the multiagent environment that is specific to transportation applications, based on space-time graphs. The chosen application example is information dissemination in disturbed transit networks. Section 4 concludes the paper and provides some future works.

2 Generic Environment Model

In dynamic transportation applications such as advanced traveler information or dial a ride systems, travelers, clients and vehicles join the system in a nondeterministic way, and might leave it anytime as well. When specifying such open systems, the designer has to define an architecture that allows for the integration of unknown agents. Newcomer agents have to be able to find the agents that have the properties, the capabilities or the resources that they need. To deal with this problem, known as the connection problem, the authors in [6] propose the concept of middle agents, who are the privileged interlocutors of agents looking for specific capacities. The author in [7] proposes recommendation systems, enabling the linking of distributed agents in open MAS. This approach allows for the progressive and distributed construction of an address book for the agents. However, in the dynamic transportation systems, the desired capabilities and sources of information are generally known: the transport operators, or the vehicles, or the real-time traffic information providers, etc. The problem is to know which of the information generated by these sources are relevant for the new agents, which context and needs are usually continuously changing. The multi-agent environment is also used for agents matching based on the properties of agents and the exchanged objects [8].

We adopt this environment-centered approach, because it focuses on the shared data and allows for the selection of relevant information without having to know or to maintain knowledge about the emitters of these data. We propose a generic representation of the environment, shared by all the agents of the system, that allows for the associative discovery of the other agents and the exchange of information between them. Agents do not maintain an address book of each others and delegate the matching of their preferences with others properties to the environment. They also can describe the properties of the agents they want to interact with and the messages they want to receive. The presence of a shared environment and the possibility to define complex interaction constraints makes this model an excellent candidate for the design of open and dynamic transportation systems.

2.1 Model

The Fig. 1 illustrates the architecture of a multiagent system (MAS) following our generic model. The modeled MAS executes on a host, where (local) agents add, read and take objects to/from the environment. Every agent is either independent (like agent 1), or representing a non-modeled external system/user in the MAS (like agents 2 and 3).

For the specification of agent behavior, we adopt four primitives inspired by Linda [9] and a set of operators borrowed from Milner's CCS [10].

A MAS written following our generic model is defined by a dynamic set of *agents* interacting with an *environment* - denoted Ω_{ENV}, which is composed of a dynamic set of *objects*. Agents can *perceive* (read only) and/or *receive* (read and take) objects from the environment. Agents are defined by a behavior (a process), a state and a local memory in which they store the data they perceive or receive from the environment. The primitives allowing these actions are the following:

$$\mu ::= add(sds) \mid spawn(P, sds) \mid look(sds_p, sds_r, e) \mid update(sds)$$

The primitive $add(sds)$ adds to the environment an object described by sds. For instance, $add(position \leftarrow 1)$ adds the property-value pair $(position \leftarrow 1)$

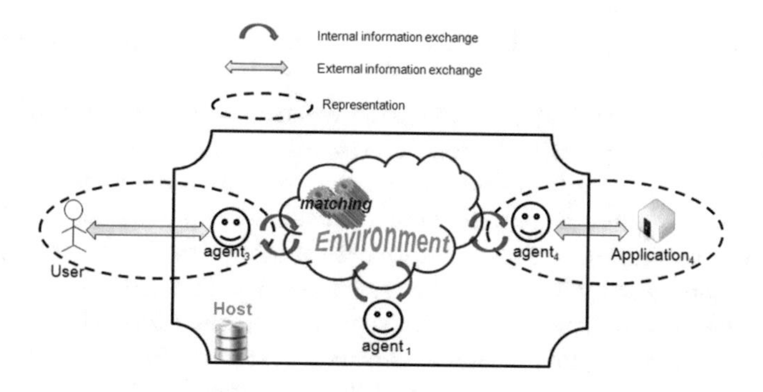

Fig. 1. Architecture of the MAS

to Ω_{ENV}. The primitive $spawn(P, sds)$ launches a new agent that behaves like P and which state is described by a description sds. For instance, $spawn(add(position \leftarrow 1), \{id \leftarrow a_1, position \leftarrow 1\})$ creates an agent that has a_1 as id and 1 as $position$ and whose behavior is $add(position \leftarrow 1)$. The primitive $look(sds_p, sds_r, e)$ allows both object perception and reception (perception and removal from the environment). It blocks until a set of objects becomes present in Ω_{ENV} such that the expression e is evaluated to $true$; the objects associated with the variables in sds_p are perceived and those associated with the variables in sds_r are received. For instance, the following instruction:

$$look(\{ticket \leftarrow t\}, \{paper \leftarrow p\}, t.destination = \text{“}Berlin\text{”} \wedge t.price \leq budget$$

$$\wedge p.decision = \text{“}accepted\text{”})$$

looks for two objects that will be associated with t and p. The object associated with t will be perceived while the object associated with p will be received. After the execution of this instruction, the two objects will be present in the local memory of the caller agent. The latter will have two additional properties: *ticket* that refers to the object associated with the variable t and *paper* that refers to the object associated with p. The perceived *ticket* has “Berlin” as destination and a *price* lower than the budget of the executing agent, while the *received* paper is “accepted” (the property *decision* is equal to “accepted”).

2.2 Environment-Centered System for Traveler Information

In this section, we describe an application based on our model. We modeled and implemented a traveler information server. The purpose of the server is to inform online travelers about the status of the parts of the transportation network that concerns them. Transportation Web services are represented with agents in the server and their properties are related to the service or the information that they provide. The problem in this kind of applications concerns the information flows that are dynamic and asynchronous. Indeed, each information source is hypothetically relevant. An agent cannot know *a priori* which information will interest him, since this depends on his own context, which changes during execution.

The objective in this application is to ensure the information of a traveler about his ongoing trip (disturbances, alerts, alternative itinerary). This process is difficult because the information sources are distributed and the management of the followup assumes a comparison of all the available information. Using our model of the environment for this application allows to design an information server parameterized by its users (the travelers). We have defined two categories of agents, the first concerns the agents representing the users (that we call PTA for Personal Travel Agent) while the second concerns the agents representing the transportation services (that we call Service Agent).

We have implemented a multiagent system running on a Web server for traveler information, where each Web service has a representant in the multiagent environment, which is responsible of the convey of messages from the server to the transportation Web service and conversely. Every user is physically mobile and connects to the server via a transportation assistant app (TAA) and has during his connection a PTA agent representing him inside the server, which is his interlocutor during his session. The context of the example is the following: inside the system, there is an agent representing a trip planning service and an agent representing a traffic service responsible of the emission of messages related to incidents, traffic jams, etc. These agents are persistent, since they are constantly in relation with the system providing the service. On the contrary, PTA agents representing the TAA in the system are volatile, created on the connection of a user and erased at the end of his session i.e. when he arrives at destination.

Every stop of the network is described by a line number *line* to which it belongs, and a number *number* reflecting his position on the line. A user u is also described by his current position in the network (the properties *line* and *number*). In a basic execution scenario, u has a path to follow during his trip i.e. a sequence of tuples $\{(line, number_{source}, number_{destination})_i \mid i \in I\}$, with I the number of transportation means used by the traveler. Every tuple represents a part of the trip, without transfer. To receive his plan, the TAA connects to the information server, and the agent u representing him is created. Then, the user is asked to specify his departure as well as his destination. Once these information entered, u adds his planning demand in the environment. A demand is an object described by his properties: *emitter*, *subject*, etc. Afterwards, u keeps on listening to messages that are addressed to him, this way: $look(\emptyset, \{message \leftarrow x\}, x.receiver = id)$. The agent representing the trip planning service is listening to messages asking for a plan: $look(\emptyset, \{request \leftarrow x\}, x.subject = \text{“}plan\text{”})$. As soon as he receives the message, he creates a message addressed to the trip planning Web service and awaits for the response. When he receives the answer, a message is added to the environment addressed to u with the received plan as body: $add(\{emitter \leftarrow id, receiver \leftarrow request.emitter, body \leftarrow plan\})$. The agent u, when he receives the message, analyzes it and displays the result on the user's TAA. Then, the agent u restrains his interaction to the messages concerning events coming up on his way. To do so, he executes the following action:

$$look(\emptyset, \{event \leftarrow x\}, \{x.subject = \text{"}alert\text{"}\}, x.line = line \wedge x.number \geq number)$$

The agent u is interested by the alerts concerning his transportation plan, which are expressed by the preceding *look* action. Let us assume that the agent representing the alert service adds an alert message concerning an accident on the way of u resulting on a serious delay for him. The traveler, via his representing agent u, is notified concerning this alert event. Since the properties *line* and *number* are updated (with an *update* action) at each move of u (each time he moves from stop to stop), the segment concerned by the alert messages gets gradually reduced until the end of the trip. The use of the environment, the constant update of the properties of the PTA agents, together with the use of *look* actions allowed us to maintain a constant awareness of the traveler about problems occurring during his trip, without relying on continuous requests to the server.

The proposal of an environment-centered system for traveler information shows how our model allows for the design and implementation of a dynamic and open transportation system. Agents join and leave the system freely and have complex interaction constraints. In this application, the interaction constraints concern the current positions and the itineraries of the travelers.

3 Space-Time Environment for Traveler Information

In this section, we present an approach where the notion of multiagent environment is used in a different way. The environment model presented above is useful for transportation applications, but remains general-purpose, and can be used for any open MAS where interaction is complex and involves several agents at the same time. The environment model that we present in this section is directly inspired and usable for transportation applications, or at least for applications involving mobile entities on a graph, a grid or a plane. The general idea is to propose a space-time representation of the environment, which can either be used as a mental model of the agents or to synchronize agents actions and movements. This representation has been used in the past in different applications: dial a ride, vehicle routing, etc. [11–13].

3.1 Generic Space-Time Model of the Environment

Given a transportation network $G = (V, E)$, with a set of nodes $V = \{(v_i)\}, i = \{0, \ldots, N\}$ and a set of arcs $E = \{(v_i, v_j) | v_i \in V, v_j \in V, v_i \neq v_j\}$. Let two matrices $D = \{(d_{ij})\}$ and $T = \{(t_{ij})\}$ of costs, of dimensions $N \times N$ (the arc (v_i, v_j) has a distance of d_{ij} and a travel time of t_{ij}). The representation of the multiagent environment is made of a duplication of G, H times, with H the maximum allowed time of the considered application: $G(t) = (N(t), E(t))$, with $N(t)$ a set of nodes at time t and $E(t)$ a set of directed links at time t, and $0 \leq t \leq H$. The time copies of G are not necessarily identical (cf. Fig. 2). Indeed, we could have different travel times between two copies to reflect the traffic status. Some nodes can be present in one copy while absent in another to reflect the expansion of a crisis situation. Arcs also can be absent to reflect vehicles timetables in public transport as in the application described in the following section.

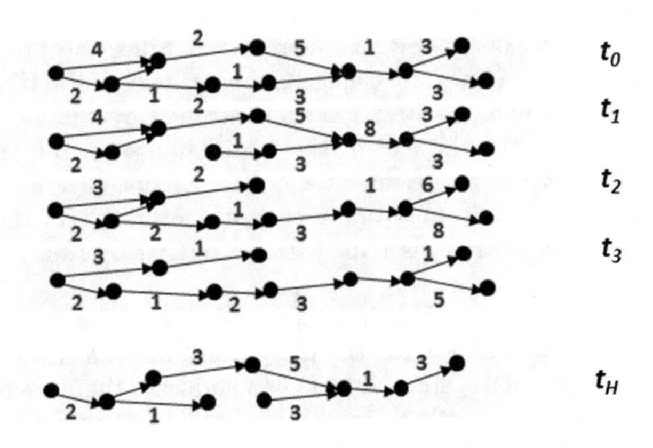

Fig. 2. Space-time network

3.2 Impact of Real-Time Information in Disturbed Transit Networks

Transportation systems are becoming progressively complex as they are increasingly composed of smart and mobile entities. Indeed, passengers mobile devices and connected vehicles allow passengers and vehicles to have up-to-date information and their behavior is now related to these information. However, without control, the massive spread of information via billboards, radio announcements and individual guidance may have perverse effects and create new traffic jams. Indeed, with this generalization of real-time traveler information, the behavior of modern transportation networks becomes harder to analyze and to predict. It is then important to observe these effects to consider the proper methods to deal with them. To this end, we have developed a multiagent simulation platform [14] that represents travelers, drivers and public transportation vehicles and make them evolve in a realistic way on a multimodal transportation network. To allow travelers to receive the only disturbance information that concerns them, we use a space-time network. In the next section, we briefly present the multiagent simulator before presenting our method for information dissemination to the relevant travelers with space-time graphs.

The multiagent simulation platform allows for the individual monitoring of travelers on a transport network. We enrich it with traveler information capabilities, both at the stops and with personal information. The simulator represents

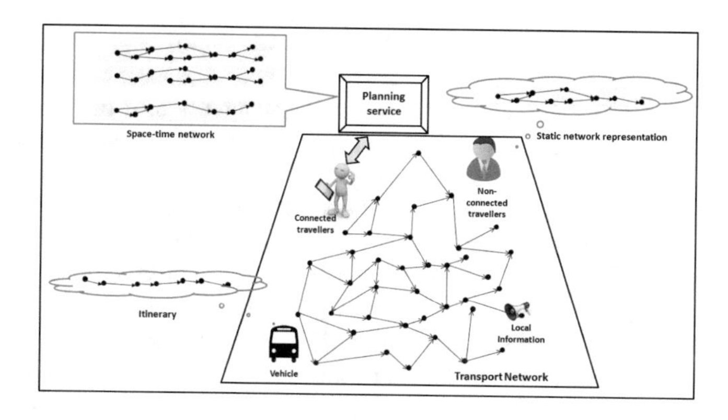

Fig. 3. Multiagent system

itinerary planners, passengers, public transportation vehicles and information means in a micro-level and simulate their dynamic movements (cf. Fig. 3).

The multiagent system of the simulator is composed of the following entities (cf. Fig. 3). The planning service has the responsibility of computing the best itinerary for the traveler agents. The planning service bases his computation on the latest status of the networks. Each active agent has a list of coordinates that he has to follow, resulting from the itinerary that he received from the planning service. At each simulation tick, the agent iteratively move from his current coordinate to the next in his list. The public transportation vehicle agents don't choose their origin and destination and obey to predefined timetables. When the vehicle reaches a stop, he looks in his onboard travelers who has to leave at this stop. Then the vehicle agent looks in the list of waiting travelers at the stop who has to take him. When they are not walking, traveler agents do not travel on their own, but take public transportation vehicles, which are responsible of their movements. The traveler agent alternates between walking and waiting for a vehicle. Local information agents represent devices that broadcast trafic information on screens or voice announcement at the stops. Every traveler, when he passes by this node at the planned time will get an update about the disturbances on the network.

Travelers that are not connected to a real-time information source base their calculation on a static view of the network. Once they have received an itinerary from the planning service, they are on their own. They will wait for vehicles at the planned stops and will not change their itinerary until they either get stuck in a disturbance (delay or line disconnection) or they receive a local information from the environment about a disturbance. When they receive the information, they will infer the new status of the network by applying the modifications to their mental - and static - view of the transportation network, and compute a shortest path based on that representation.

We model the public transportation network as a space-time environment, representing both the network topology and the vehicles timetables. An arc connects two nodes $n_1(t)$ and $n_2(t)$ in $G(t)$ when there is a vehicle departing from n_1 at t. Otherwise, the arc is absent. Space-time arcs are active in this application: they store listeners from the traveler agents and inform them when the departure time or the travel time changes or when the mission of the vehicle is deleted. It is necessary to circumscribe the broadcast of messages to nearby traveler agents. At each concerned node of the space-time network, an information device is associated. To be aware of the only information that concern their stop, local information agents subscribe to space-time edges connected to the concerned stop. To be aware of the only events that concerns him, the connected traveler agents subscribe to the only space-time arcs of the multiagent space-time environment that form his itinerary. When the travel time of an arc or the departure time of the vehicle changes, the information is broadcasted to the subscribing connected travelers. The planning process is then launched with the new status of the network.

Disturbances are modeled exclusively by modifying space-time arcs (that correspond to vehicles timetables). Indeed, delays are injected in the model by dynamically modifying the timetables of the vehicles, adding some time to arrival times. Breakdowns are modeled also by deleting a part of the mission of a vehicle. To model the breakdown of an entire line, the timetables of the remaining vehicles of the line are all deleted. As soon as a timetable is modified, based on the space-time network, the information is immediately detected by the concerned local information devices at stops. The concerned connected travelers will also receive the information immediately. Hence, when a timetable is modified, the information about the delay or the breakdown is sent to the only connected travelers that are interested by these vehicles missions.

4 Conclusion

This paper is based on the conviction that multiagent systems are a suitable paradigm for modeling, simulating and optimizing dynamic transportation applications. It investigates one research question: the explicit modeling of the multiagent environment is it a good choice for these applications? The answer suggested by our work is yes, and we propose two classes of environment models that are interesting for the design of transportation applications. The first class concerns generic environment models for interaction. The second class proposes a space-time model for interaction and is supported by a space-time representation of the environment.

The design of the multiagent environment as an explicit entity is often criticized because it introduces centrality in systems that are supposed to be completely distributed. Following these arguments, centrality could lead to communication bottlenecks, to weak fault tolerance and to poor scalability [15]. However, as we can see it in the models and applications presented in this paper, this architecture has several benefits, and we believe that there is a compromise between the two visions. In our ongoing work, we develop the idea that we still can benefit from an explicit representation of the multiagent environment without loosing the benefits of distribution, namely fault tolerance and scalability.

This is done by splitting the design process in two phases. During the first phase, the system is designed with a conceptually centralized environment. During the second phase, the multiagent environment is distributed. We are working on environment distributions for each type of environments presented in this paper.

References

1. Bazzan, A.L., Klügl, F.: A review on agent-based technology for traffic and transportation. Knowl. Eng. Rev. **29**(03), 375–403 (2014)
2. Bessghaier, N., Zargayouna, M., Balbo, F.: Management of urban parking: an agent-based approach. In: Ramsay, A., Agre, G. (eds.) AIMSA 2012. LNCS, vol. 7557, pp. 276–285. Springer, Heidelberg (2012). doi:10.1007/978-3-642-33185-5_31
3. Bessghaier, N., Zargayouna, M., Balbo, F.: An agent-based community to manage urban parking. Adv. Intell. Soft Comput. **155**, 17–22 (2012)
4. Parunak, H.V.D., Savit, R., Riolo, R.L.: Agent-based modelling vs. equation-based modelling: a case study and users' guide. In: Proceedings of Workshop on Modelling Agent Based Systems (MABS 1998), Paris (1998)
5. Weyns, D., Michel, F. (eds.): E4MAS 2014. LNCS (LNAI), vol. 9068. Springer, Cham (2015)
6. Sycara, K., Wong, H.: A taxonomy of middle-agents for the internet. In: Proceedings of the Fourth International Conference on MultiAgent Systems (ICMAS-2000), Washington, DC, USA. IEEE Computer Society, pp. 465–466 (2000)
7. Vercouter, L.: Conception et mise en oeuvre de systèmes multi-agents ouverts et distribués. Ph.D. thesis, Ecole Nationale Supérieure des Mines de Saint-Etienne, Université Jean Monnet-Saint-Etienne (2000)
8. Zargayouna, M., Trassy, J.S., Balbo, F.: Property based coordination. In: Euzenat, J., Domingue, J. (eds.) AIMSA 2006. LNCS, vol. 4183, pp. 3–12. Springer, Heidelberg (2006). doi:10.1007/11861461_3
9. Gelernter, D.: Generative communication in linda. ACM Trans. Program. Lang. Syst. **7**(1), 80–112 (1985)
10. Milner, R.: Communication and Concurrency. Prentice-Hall, Upper Saddle River (1989)
11. Zargayouna, M.: Une représentation spatio-temporelle de l'environnement pour le transport à la demande. In: Représentation et raisonnement sur le temps et l'espace, Nice, France (2005)
12. Zargayouna, M., Zeddini, B.: Fleet organization models for online vehicle routing problems. In: Nguyen, N.T. (ed.) Transactions on Computational Collective Intelligence VII. LNCS, vol. 7270, pp. 82–102. Springer, Heidelberg (2012). doi:10.1007/978-3-642-32066-8_4
13. Zargayouna, M., Balbo, F., Scemama, G.: A multi-agent approach for the dynamic VRPTW. In: Proceedings of the International Workshop on Engineering Societies in the Agents World (ESAW 2008), Saint-Etienne. Springer (2008)
14. Zargayouna, M., Zeddini, B., Scemama, G., Othman, A.: Agent-based simulator for travelers multimodal mobility. Front. Artif. Intell. Appl. **252**, 81–90 (2013)
15. Billhardt, H., Fernández, A., Lujak, M., Ossowski, S., Julián, V., Paz, J.F., Hernández, J.Z.: Towards smart open dynamic fleets. In: Rovatsos, M., Vouros, G., Julian, V. (eds.) EUMAS/AT -2015. LNCS, vol. 9571, pp. 410–424. Springer, Cham (2016). doi:10.1007/978-3-319-33509-4_32

Microservices as Agents in IoT Systems

Petar Krivic(✉), Pavle Skocir, Mario Kusek, and Gordan Jezic

Internet of Things Laboratory, Faculty of Electrical Engineering and Computing, University of Zagreb, Unska 3, HR-10000 Zagreb, Croatia
{petar.krivic,pavle.skocir,mario.kusek,gordan.jezic}@fer.hr

Abstract. Developing robust monolith systems has achieved its limitations, since the implementation of changes in today's large, complex, and fast evolving systems would be too slow and inefficient. As a response to these problems, microservice architecture emerged, and quickly became a widely used solution. Such modular architecture is appropriate for distributed environment of Internet of Things (IoT) solutions. In this paper we present a solution for service management on Machine-to-Machine (M2M) devices within IoT system by using collaborative microservices. Collaboration of distributed modules highly reminds of multi-agent systems where autonomous agents also cooperate to provide services to the end-user. Because of these similarities we consider microservices as modern agents that could improve systems in distributed environments, such as IoT.

Keywords: IoT · M2M · Microservices · Agents · Service management

1 Introduction

Internet of Things (IoT) is not just a popular topic generating a lot of interesting scientific prototypes, but a very popular concept with a lot of existing industrial solutions in everyday use. Connecting a large amount of devices to Internet has shown a lot of benefits and possibilities, but also leaves space for improvement. It is important to offer users and companies interested in this domain a large number of different possibilities and architectures, so they could choose the one that suits them the most.

Common architecture of IoT system has a robust server node where data is processed and stored. It is also connected with M2M devices that generate this data through M2M gateway, a middle-node that translates messages of different protocols. It turned out that processing such amount of data could cause problems on standard servers and alternative was needed. Processing data in cloud is today the best alternative, but other options should be studied as well, so that producers have more options to combine and apply to their products, depending on their needs. Because of the distributed environment of IoT systems we believe that using microservices in their implementation would be an applicable approach, since they also have distributed architecture.

G. Jezic et al. (eds.), *Agent and Multi-Agent Systems: Technology and Applications*, Smart Innovation, Systems and Technologies 74,
DOI 10.1007/978-3-319-59394-4_3

Lately, microservices became a very popular topic in software development. Since it was hard to maintain vast amounts of code in large companies, an idea of modularization proved to be a good alternative. Modular code is easier to upgrade, to fix if needed and new employees can get familiarized with it quickly. Also, teams could focus on parts of the system that are their responsibility without the need of closely knowing the code of the entire system. Microservice architecture has a lot of similarities with IoT systems [3], which was our motivation to implement the system described in continuation of this paper.

Distributed microservice architecture has also similarities to agent system architecture. Requests are processed collaboratively and only the end result is delivered to end-user. In order to test the behavior of microservices in IoT system, and to point out similarities between them and agents in multi-agent systems, we present a solution described in continuation.

Section 2 gives a short overview of related work in connection with IoT, microservices and agent-based systems. Section 3 presents the purpose of service management in our system and proposed architecture used to achieve it. Next, Sect. 4 describes advantages of microservice architecture and a comparison between agents and microservices is given. Finally, Sect. 5 concludes the paper.

2 Related Work

IoT systems architecture was inherited from Machine-to-Machine (M2M) concept. Since this concept is slightly older then Internet of Things there are already some existing standards in use [4] where architecture is proposed, and a great number of prototypes has already accepted it. Bhowmik et al. [2] used proposed architecture in their project intended for ambient assisted living (AAL). They connected constrained devices that perform measurements with their social network server through gateway. This is the most common case of data processing in M2M applications, where gateway forwards sensor data to server continuously.

Repetitive measurements of sensor values and forwarding them to gateway can be challenging for battery-powered devices with limited energy available. Finding more efficient solution of this problem is also a very popular research topic. In [9] a solution where interval of measurements would increase if observation values were similar is proposed, assuming that changes in observed space are not as frequent as expected. This way the amount of measurements and sending data is reduced, thereby accomplishing energy savings.

Internet of Things architecture is quite similar to microservice architecture as Butzin et al. conclude in [3]. They compared distributed architecture of IoT systems with microservices, and pointed out that both of them have the same architecture goal. Since they are both modular, their goal is to create application which unifies a large number of services. Main differences between these two are a result of processing limitations that M2M devices have. This is the reason why in some cases microservice architecture is easier to deploy and manage then it is the case with IoT solutions.

Distributed architecture of microservices suddenly became widespread, but new challenges emerged with it. In [10] Xu et al. highlight programming systems that consist of a large number of services, reducing communication overhead in these systems and flexible deployment of services to a network as three main challenges in microservice system construction. In their work they tackle these challenges with agent-oriented programming language CAOPLE. They propose separating service as functionality from service as a computational entity that offers such functionalities. So in their novel language service is functionality offered by agent, where they point out similarity with object oriented design. Microservices in their work are then presented as agents which can communicate using some of the provided mechanisms. Also, in [5] Liu et al. present a software development environment where programming in previously mentioned CAOPLE is enabled and the user can test the execution of created application.

In this paper we demonstrate the similarities between agents and microservices. Firstly, we describe our goal with two approaches we used to achieve it. Then a comparison between two used approaches is given with a review on their compliance with popular software architecture properties. Finally, we describe similarities of microservices and agents with a conclusion why we think there is a significant resemblance between these two concepts.

3 Agent Model for Service Management in IoT Systems

In order to test agent network model based on microservices, we decided to implement a solution which would enable service management on M2M devices in our M2M network. The main goal was to achieve two-way communication with M2M devices which is then used in registration process of devices and services in our system, and to start or stop registered services based on the requirements of interested parties.

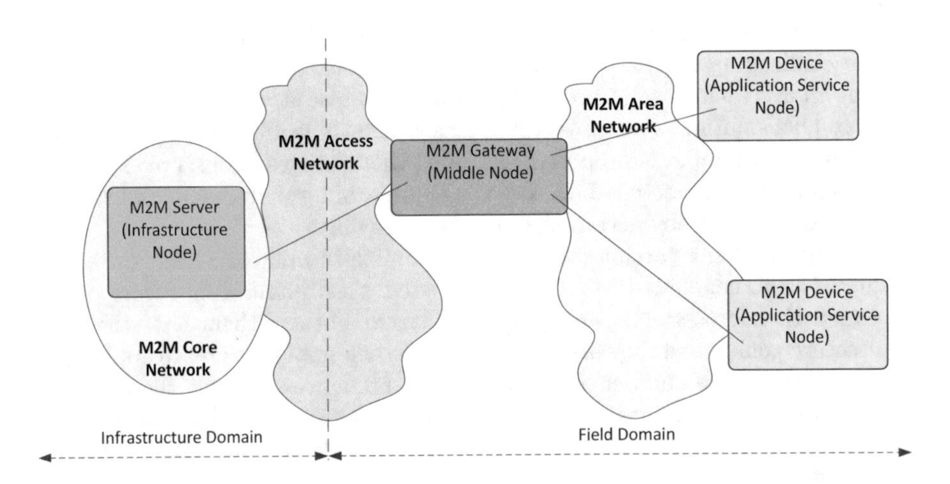

Fig. 1. oneM2M architecture

OneM2M's recommended architecture [7] for M2M systems depicted in Fig. 1 shows the environment of our solution. We used Waspmote PRO v1.2 as our M2M devices which communicate with Waspmote gateway using the XBee 802.15.4 area network. Since our goal was to enable communication between the end-user accessing our system over the Internet and M2M devices in our M2M area network, we had to implement and integrate agents in both of previously mentioned networks. In M2M area network we increased functionalities of M2M devices to make them more agent oriented, which is described in the following subsection.

However, the main focus of this paper is to present agent model in IoT system based on microservices, used for implementation of M2M server and M2M gateway in Fig. 1. Gateway is a middle node that has a role to forward messages between M2M server and M2M devices. Our M2M server consists of user-interface where user can manage services, database where all data is stored and communication interface towards gateway. Also, to compare advantages and disadvantages of proposed microservice model with other approaches, we developed a monolithic model as well. More detail description of these two models is given in continuation.

3.1 Agent Based M2M Devices

Since Internet of Things concept emerged, usual practice was to collect data from M2M devices continuously, which is not an energy efficient approach. Furthermore, in this way a lot of redundant or unnecessary data is measured, sent to network and stored in databases [1].

This could be resolved, or at least improved, if users could somehow manage existing services on M2M devices. In order to achieve this goal we extended mentioned usual practice so that our M2M devices could adapt to the end-user requirements. The idea was to create a system that could show all available services offered by reachable M2M devices. The user could then start or stop selected service, and set the time frequency of measurements in order to spare energy, but also to receive requested data in specified time intervals.

Algorithm 1 shows proposed procedure for M2M devices in M2M network that are a part of our system. Firstly, the device waits for messages from gateway, scheduled at the time of the wakeup. There are two types of messages it can receive: request for registration of services offered by device and request to start or stop the specified service. If the registration request is received, registration data is sent towards gateway. Otherwise, if start or stop request is received, device turns on or off (depending on the current state) measurements for the specified service. Also, in this case device updates interval of hibernation to the received value. Afterwards, the device performs measurements and sends the collected data if any service is active. Finally, the device goes to sleep for specified amount of time, after which the process is repeated.

Algorithm 1. Proposed M2M device algorithm

```
1: procedure TWO-WAY COMMUNICATING DEVICE
2: loop:
3:     receive messages from gateway.
4:     if message received then
5:         if registration request received then
6:             send device registration data.
7:         else
8:             if measuring and sending == ON then
9:                 measuring and sending = OFF
10:            else
11:                measuring and sending = ON
12:            x = received time interval.
13:    if measuring and sending == ON then
14:        measure sensor values.
15:        send measured sensor values.
16:    sleep (x).
17:    goto loop.
```

3.2 Architectures Description

In our first approach we implemented a monolith M2M server that communicates using the HTTP communication protocol with M2M gateway. Our goal was to compare microservices approach with monolith architecture but because of the distributed environment in IoT, we had to develop two-component system. First component consists of data manipulation, application logic and interaction with the end-user and it is the monolithic core of our system. Second component is the gateway that forwards messages between M2M network and the mentioned core. Since in this case gateway is just a necessary separate part of the our core we decided to implement it as a simple runnable application. Thus, it can be deployed on every device which has a gateway for XBee 802.15.4 communication attached to it.

To achieve all necessary functionalities of our system using microservice architecture, we developed three microservices (Fig. 2). Two of them form the M2M server and the third is M2M gateway in Fig. 1. Since it is recommended that one microservice should be small and autonomous but also collaborative [6], we designed mentioned architecture where every microservice performs simple tasks, but in collaboration with other microservices all functionalities of our system are achieved. It is also recommended that each microservice should have its own business value. However, in distributed environment such as IoT this is hard to achieve, so in our system they offer business value only in collaboration. Communication between microservices is accomplished using the HTTP protocol, so each microservice has its own RESTful interface. Tasks of each microservice are:

- Microservice 1 (Database agent) - Stores data received from gateway to database;

- Microservice 2 (Gateway agent) - This microservice has several tasks since it communicates with 802.15.4 network. Firstly it registers itself to database agent, so database agent knows the address where gateway is available. Then, it periodically discovers reachable devices in 802.15.4 network and initiates their registration. Finally, this agent forwards all data received from M2M devices towards database agent and the other way around.
- Microservice 3 (User Agent) - This agent interacts with the end-user. It offers interface where all available devices are shown and then the end-user can manage offered services.

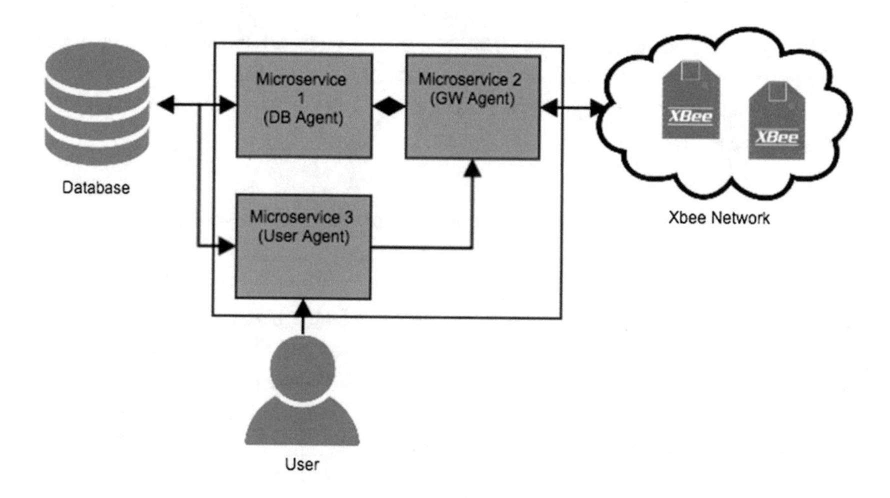

Fig. 2. Microservice architecture

4 Benefits of Microservice Approach in IoT Systems

As mentioned before, we implemented a system that enables service control on M2M devices in M2M network using both of the previously described approaches. We achieved two-way communication between the end-user and M2M devices. Consequently we achieved our goal in energy savings on M2M devices and also, we reduced the need of physical interaction between system provider and mentioned devices. System provider could now control active devices remotely which is important in cases when devices are physically hard to reach.

In continuation we present our comparison of monolith and microservice architecture, with more detail description why we propose microservice architecture over others in research oriented systems. Also, we describe how each microservice could in fact be considered as a single agent. System made of several microservices is then a multi-agent system where each agent, in collaboration with others, processes requests.

4.1 Comparison of Microservices with Monolith Approach

The greatest advantage that microservice architecture offers in comparison to other architectures is agility. One of the motives for creating microservices was to modularize complex and robust systems, in order to increase speed of deploying new solutions or upgrading the old ones [6].

We also first developed monolith system to solve previously described problem, and then decided to modularize it. Our motivation to do so were constant changes and improvements we had to do on our system, since we wanted to research different possibilities in the IoT domain. Every change would then impact the whole system and consequently we had to revise the whole code while the system would be unavailable. So in order to avoid this we decided to use microservices in future, which is suitable for research and development projects where a lot of changes are made daily, and the scope of the developing system is always increasing.

Table 1. Property comparison for each architecture

Property	Microservice architecture	Monolith architecture
Scalability	Replicating microservices across servers depending on demand	Replicating the whole monolith system on multiple servers
Development	Different programming languages in one system	All services written in same programming language
Changeability	Changing occurs on a single microservice	Changing one service causes revisioning and redeploying the entire system
Integration	Challenging because of the large number of distributed modules	Modules placed together
Communication	Defined communication interface between microservices	In-memory
Maintenance	Smaller pieces of code	One source code for the whole system
Upgrades	Adding new microservices	Upgrading the whole system

Table 1 shows some of the most popular properties that are considered when designing a new system. Scalability is easier to achieve when using microservice architecture since the owner has to replicate only microservices that are overloaded. In case of monolith architecture the owner has to replicate the complete system on multiple servers and add a strategy for load balancing between system replicas. Development is also more convenient with microservices. System architect can have heterogeneous development team regarding programming languages, and still create fully functioning solution. Every microservice can be implemented in different programming language as it is only important that the communication interface between them is well defined. Again, in monolith architecture everything should be implemented using the same programming language. Agility and upgrades are also much easier in modular systems because changing small modules or creating and integrating new ones is much easier when they do not impact the complete system. If a single microservice has a key role in the system, it can be easily replicated and changed when the new one is compliant with communication interface used in the system. Equally, maintenance of smaller code pieces that implement modules is easier then updating robust monolithic source code.

However, monolith architecture has also its advantages as for example communication between services and integration of the system. Since all services are a part of the same system placed on a single server, they communicate using the same memory and their integration is implied. In our opinion communication could be the greatest weakness of microservice architecture, since a great amount of communication between different services occurs on every request. In most cases HTTP is not efficient enough, and alternatives are different messaging systems such as RabbitMQ, Apache Kafka etc. These messaging systems accelerate message exchange between microservices, but since in most cases there is no possibility of parallel processing there is still time overhead compared to monolith system procedure time.

4.2 Microservices as Agents

Because of the distributed architecture and other properties, microservices are in our opinion modern agents. As it can be seen in Table 2, microservice has almost every well known agent property [8]. For that reason we could present microservice architecture of our system as a collaborative multi-agent composition. Every microservice in our system is then in fact an agent which performs modest tasks, and in collaboration with other agents it delivers a fully functional service to the end-user.

Autonomy of every microservice is obvious since it can accomplish its task without human intervention and independently of other services. In Fig. 2 all three microservices could in fact be run as a stand-alone services. Microservice 2 for example has a gateway role. Once it is up, it forwards messages and performs other activities completely autonomously. Microservice 1 stores data in database, and Microservice 3 fetches data from it, so it could offer GUI to the end-user. Therefore, these could equally work autonomously.

Interaction, collaboration and coordination are three basic properties of microservices. A modular system consisting of multiple microservices must have a well defined interface between them, in order to work properly and efficiently. Every microservice in Fig. 2 would not have much purpose on its own, but together they form a complete well functioning IoT system. For example if a user wants to start a service and receive data from specified device, it can send request using Microservice 3. However, if Microservice 2 does not forward request to the device and receive data from it, or on the other hand, if Microservice 1 does not store data to database, user will not receive the wanted result.

Mobility is also easy to achieve by using microservices. System owner can run and move each microservice independently, as long as other microservices can reach it. If HTTP protocol is used for communication between them, they must have an option to update the address of moving agent. In case messaging system is used, this is also not necessary since then the moving agent only needs to maintain or establish new connection towards messaging broker.

Reactivity of microservice agents in our system is defined as their implementation logic. Once our microservices are up and running they must be aware of the data processing procedure. Also, if we consider supporting additional communication protocols on gateway (Microservice 2), reactivity is even more highlighted. In that case agent is a multi-protocol reactive bridge of the system, which is responsible for using correct procedures in the right moment.

Learning and adaptation are two agent properties which are not contained in microservices of our system. However, these properties are also achievable in microservices. They are left out from our system for now, but in future upgrades they might be considered.

Table 2. Software agent properties in microservices

Property	
Autonomy	✓
Adaptation	✗
Interaction	✓
Mobility	✓
Learning	✗
Collaboration	✓
Coordination	✓
Reactivity	✓

5 Conclusion

This paper presented benefits of microservice architecture and its similarities with agent-based architecture. In our opinion the greatest improvement brought by microservices is agility of system development. After modularizing our monolith system and shifting it to microservice architecture we could focus on parts we wanted to research, without the need of revising the whole code after every new feature. Since each microservice is independent individual that performs a single service, we presented them as agents in our systems. Each agent has its own purpose and together they form a collaboration that accomplishes the goal of the system.

In our future work we plan to implement additional agents in order to offer new services to end-users. We also plan to add learning and adoption properties to our microservices to achieve dynamic adjustments to the environment they are placed in. Also, in this way our microservices will have all the properties of the agents. Therefore we think this will be indisputable argument that in future microservices should be used to implement agent-based systems.

Acknowledgements. This work has been supported by Croatian Science Foundation under the projects 8065 (Human-centric Communications in Smart Networks) and 8813 (Managing Trust and Coordinating Interactions in Smart Networks of People, Machines and Organizations), and by Croatian Regulatory Authority for Network Industries under the project Looking to the Future 2020.

References

1. Alduais, N.A.M., Abdullah, J., Jamil, A., Audah, L.: An efficient data collection and dissemination for IOT based WSN. In: 2016 IEEE 7th Annual Information Technology, Electronics and Mobile Communication Conference (IEMCON), pp. 1–6 (Oct 2016)
2. Bhowmik, A.K., Khendek, F., Hormati, M., Glitho, R.: An architecture for M2M enabled social networks. In: 2015 14th Annual Mediterranean Ad Hoc Networking Workshop (MED-HOC-NET), pp. 1–8, June 2015
3. Butzin, B., Golatowski, F., Timmermann, D.: Microservices approach for the internet of things. In: 2016 IEEE 21st International Conference on Emerging Technologies and Factory Automation (ETFA), pp. 1–6, September 2016
4. ETSI: ETSI M2M solution introduction. About (2014). http://www.etsi.org/images/files/Events/2014/201405_DGCONNECT_SmartM2MAppliances/ETSI_M2M_introduction_main.pdf
5. Liu, D., Zhu, H., Xu, C., Bayley, I., Lightfoot, D., Green, M., Marshall, P.: CIDE: an integrated development environment for microservices. In: 2016 IEEE International Conference on Services Computing (SCC), pp. 808–812, June 2016
6. Nadareishvili, I., Mitra, R., McLarty, M., Amundsen, M.: Microservice Architecture, 1st edn. O'Reilly Media, Inc., Sebastopol (2016)
7. oneM2M: M2M Functional Architecture. Technical specification (2015). http://www.onem2m.org/images/files/deliverables/TS-0001-Functional_Architecture-V1_6_1.pdf
8. Pandey, A.K., Vasishtha, A.K., Saxena, A.S.: Properties and interaction of object oriented software agent with system. In: 2016 3rd International Conference on Computing for Sustainable Global Development (INDIACom), pp. 1141–1143, March 2016
9. Skocir, P., Maracic, H., Kusek, M., Jezic, G.: Data filtering in context-aware multi-agent system for machine-to-machine communication. In: Jezic, G., Howlett, R.J., Jain, L.C. (eds.) Agent and Multi-Agent Systems: Technologies and Applications. SIST, vol. 38, pp. 41–51. Springer, Cham (2015). doi:10.1007/978-3-319-19728-9_4
10. Xu, C., Zhu, H., Bayley, I., Lightfoot, D., Green, M., Marshall, P.: CAOPLE: a programming language for microservices SAAS. In: 2016 IEEE Symposium on Service-Oriented System Engineering (SOSE), pp. 34–43, March 2016

Enhancing Tactical Information Assessment Using an Agent-Based Cognitive Architecture

Angela Consoli(✉)

Defence, Science and Technology Group, Department of Defence,
PO Box 1500, Edinburgh, SA 5111, Australia
angela.consoli@dsto.defence.gov.au

Abstract. The actualisation of an information-rich, expansive and complex battlespace have rendered current Tactical Information Assessment (TIA) within tactical systems as non-effective. Existing TIA techniques and algorithms perform limited cognitive processing because they lack fundamental characteristics of Situation Management (SM). In consequence, a dramatic reduction of situation awareness, information and decision superiority of the operators and the overall tactical system has emerged. Considerable attention in applying computational cognitive architectures and processing and SM within TIA has been gaining momentum.

This paper discusses the Cognitive Architecture for Tactical Information Assessment (CATIA), a proposed Multi-Agent System (MAS)-based cognitive architecture. CATIA employs cognitive architecture design principles and the Belief, Desire and Intention (BDI) framework to facilitate recognition and reasoning to deliberate in tactical situations and events. CATIA will be implemented within the Future Integrated Mission System (FIMS) to illustrate how superior TIA methodologies can dramatically improve information assessment, situation awareness and information and decision superiority within tactical systems.

Keywords: Multi-agent systems · Cognitive architectures · Tactical information assessment · Situation management · Situation awareness · Belief · Desire and intention

1 Introduction

For future information-rich battlespace environments, current Tactical Information Assessment (TIA) methodologies will not be able to cope with the asymmetric nature of smart or intelligent threats. The risk of TIA's ineffectiveness could lead to lethal consequences. The primary areas of failure include data and information fusion techniques and algorithms that have not been designed or developed from a cognitive perspective, where decisions being made use a closed-world view of situational awareness [7,14,15]. The second is centred around the design philosophy of current tactical systems. Contemporary tactical systems

G. Jezic et al. (eds.), *Agent and Multi-Agent Systems: Technology and Applications*, Smart Innovation, Systems and Technologies 74,
DOI 10.1007/978-3-319-59394-4_4

have not been designed to be adaptive within their environment, and cannot support automated cognitive information processing. The result is the reduction of information and decision superiority against adversaries within the battlespace.

New generation tactical systems will need to perform in higher levels of uncertainty and complexity. To achieve tactical superiority, there will be greater scrutiny on the design of future tactical systems, particularly on how effectively information is collected, processed and assessed to facilitate decision-making. Tactical systems will need to incorporate computational cognitive processes, or a Cognitive Architecture (CA) to exhibit human perception, recognition, reasoning and decision-making.

CA's within tactical systems have been used to reduce the human cognitive load [13]. In recent times, significant research has been conducted into using CA's to enhance and potentially automate situation and threat management, assessment and representation in tactical systems. A current research activity is the development of an architecture to improve TIA. The proposed Cognitive Architecture for Tactical Information Assessment (CATIA) synergises cognitive and multi-agent system architecture design patterns and Situation Management. The proposed architecture will establish a new generation of TIA that has the capability to deliver the required decision superiority of future battlespaces.

This paper is structured as follows. Section 2 discusses the definition, requirements and designs needed for a CA. Section 3 explores the concept of TIA, Situation Management (SM) and issues surrounding current TIA practices and techniques. Section 4 introduces the Cognitive Architecture for Tactical Information Assessment (CATIA), the design considerations and the architecture. Section 5 discusses how CATIA will be integrated into the Future Integrated Mission System (FIMS), a new generation architecture for combat and mission systems, as well as the advantages of integrating a CA within mission systems. Finally Sect. 6 will examine the future work of CATIA.

2 Cognitive Architecture Definition, Requirements and Design

2.1 What Is a Cognitive Architecture (CA)?

A CA is described as a computational framework, where its primary role is to model, analyse and understand the human mind [1,2,16,21]. A CA facilitates and exposes detailed computational processes of cognition, which are multi-level and multi-domain [13,21]. Most importantly, CAs allow human-like intelligent activities to be reproduced within technology. The research into CAs leverage across many science disciplines that include Artificial Intelligence (AI), Computational Intelligence (CI), cognitive science, psychology and computer science [13,16,23].

ACT-R [1], Soar [11], ICARUS [12] and CLARION [22] are a number of notable CAs that have been developed and extensively detailed. Gluck (2010) believes each of these architectures need to be evolutionary and focus on a cognitively aligned phenomena. Taatgen and Anderson (2010) agrees and states that a CA must allow for the discovery and refinement of its mechanisms for cognition, where the designer must decide whether these cognitive mechanisms are still appropriate for incorporation within the architecture. An example of this characteristic is seen with ACT-R, Soar and CLARION. Taatgen et al. (2006) explicitly states that ACT-R is continuously being updated and expanded, Wray and Jones (2006) illustrates that a previous version of Soar required further development in its planning components as it only provided low-level planning capabilities and Sun (2006) indicates the computational cognitive components required for further evolution in CLARION.

CAs are developed based on an application or capability that will establish a greater and computational understanding into a particular cognitive phenomena [13,24]. ACT-R was developed for exploring procedural and declarative knowledge for memory and learning. Soar was developed to explore and demonstrate general intelligent behaviours within agents [11,26] and CLARION was developed to explore social simulation in and between cognitive agents [22].

2.2 Design Considerations and Requirements of a Cognitive Architecture

A well designed CA needs to be adaptable, directable, understandable and trustworthy [10]. Adaptability ensures that agents can be reactive, using behaviours to determine decisions and actions. Directability ensures the architecture appears sequential and routine for behavioural responses, understandability is also seen as synergistic interactions, enabling proactiveness, interactions and collaborations between implicit and explicit computational cognitive components. Finally, there must be trust between teams of cognitive architectures, i.e. human-to-machine, machine-to-human and machine-to-machine [10,23].

The successful integration of these capabilities relies on the implementation of fundamental computational cognitive functions, which include perception, recognition and categorisation; prediction; reasoning; planning; decision-making; and memory and learning [11]. To achieve any or all of the fundamental computational cognitive functions, the basic design requirements of a CA include a symbolic and knowledge-centric environment. A CA must have an environment that is sensory, diverse, complex and dynamic and must support interaction and sociality. Symbolic structures allow manipulation of descriptions on environmental objects, accept instructions from other participants and implicit communication. In addition, symbolic mechanisms provide the basics of representation [11].

3 Tactical Information Assessment and Situational Awareness

3.1 Tactical Information Assessment

Current tactical systems have the ability to provide sophisticated tactical information assessment that ensures operators have the most current Common

Operating Picture (COP) to make decisions. TIA is defined as the process of situational and threat assessment to achieve overall Situation Awareness. Traditionally, TIA algorithms and techniques have been developed under the guidance of the popular JDL Data Fusion Model, where Levels 2 and 3 are Situation and Threat (or Impact) Assessment [20]. Situation Assessment is described as determining an entity's association and estimates relating to their cumulative events, activities or behaviours. Threat assessment leads from situation assessment by reasoning and predicting the effects of an entity, its association and estimates. Due to the nature of situation and threat assessment, this process is described as higher-level, or information fusion [20].

3.2 Issues and Challenges in Automated TIA

The main challenge with current TIA methodologies is the vast amount of tactical information that needs to be analysed, interpreted and abstracted into knowledge and context for threat assessment and eventual decision-making [3,18,19]. The issues are amplified because situation and threat assessment needs to support asymmetric warfare at strategic, operational and tactical levels, hence requiring access to various sources of data and information [19]. To overcome these issues, automated systems have been designed and developed, however, the computational solutions have been limited. Powell (2005) believes current computational and automated situation and threat assessment have been underwhelming when required to manage and use information that may have different abstraction levels and relevance, and manipulating the information to represent, characterise, recognise and project the knowledge of the current situation [3,5]. Surmounting challenges faced by current automated TIA systems can be accomplished by exploring the concepts and characteristics of Situation Management (SM) [9,19].

3.3 Assessing, Representing and Analysing the Tactical Situation

The main attribute of any dynamic environment is the continual change in its *situation*. The ability to work within such an environment is dependent on how well the situation is managed. SM is focussed on how well a new situation is recognised, reasoned and predicted [8]. Jakobson (2013) defines SM as goal-directed processes that consist of sensing and information collecting, perception and recognition of situations, analysis of past situations, prediction of future situations and reasoning, planning and implementation of actions. Figure 1 illustrates the general view of SM.

SM, by its very nature, is cognitive, as its components exhibits perception, memory, learning and deliberation [8]. SM has three main components: situation investigation, situation control and situation prediction. Situation investigation refers to a cogitative analysis of situations to determine why they have occurred, situation control is viewed as determining whether to change or retain the situation based on the goals of the environment and situation prediction is the assessment of possible future situations [8]. Jakobson (2013) believes the most critical

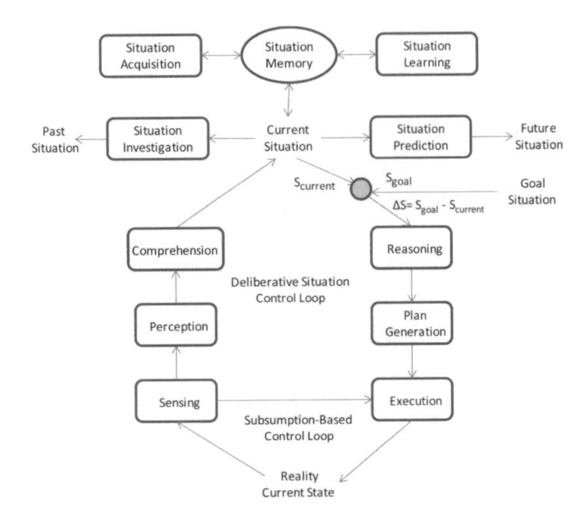

Fig. 1. Situation management view (After Jakobson (2013))

component within the SM model is situation control. The situation control loop is composed with deliberative and subsumptive-based loops, where the former is related to the cognitive processes required within a CA and the focal point of this paper. The deliberative control loop component of the SM model can be regraded as the vehicle of instantiating mental modelling and producing mental representation of the environment. Jakobson (2013) agrees and discusses how the deliberative control loop resembles the Observe-Orient-Decide-Act (OODA) loop, where the focus is providing a SM model with the sensing, perception and comprehension of entities, their states and the relations between them.

4 The Cognitive Architecture for Tactical Information Assessment (CATIA)

4.1 What Is CATIA?

The Cognitive Architecture for Tactical Information Assessment (CATIA) is a tactical CA designed to support current and future situation and threat assessment by utilising concepts, characteristics and features from situation management. As shown in Fig. 2, CATIA enables the recognition, perception and comprehension of information-rich environments, asymmetrical threats and ambiguous situations, in unison with tactical mission systems. The implementation of a tactical CA in conjunction, or within, tactical systems will enable TIA to produce far more accurate threat behaviours. The result is the tactical systems will represent far more descriptive, estimated and plausible projected actions for superior decision-making.

CATIA exploits the Belief, Desire and Intention (BDI) framework and Multi-Agent System (MAS) design patterns, which is ideal for TIA and SM [4].

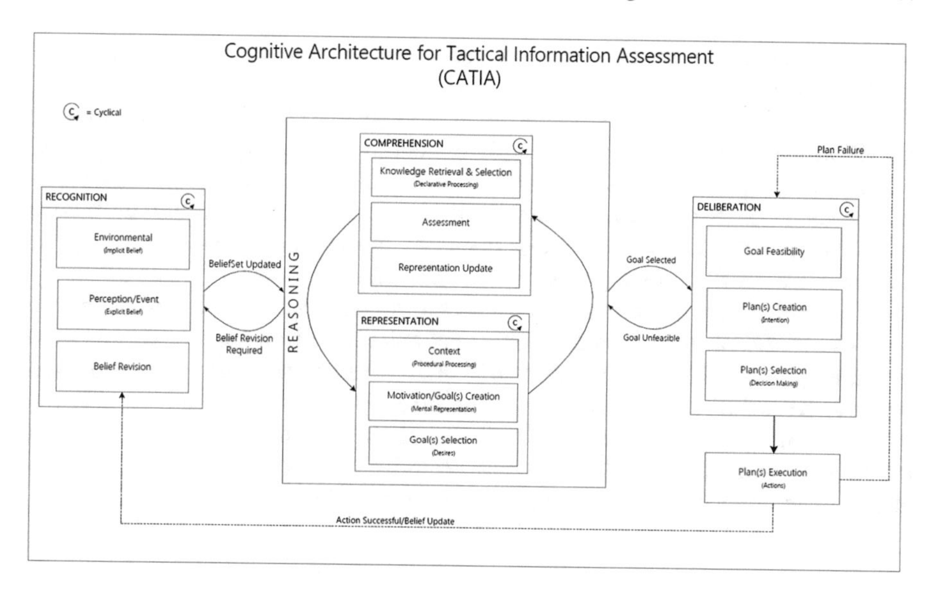

Fig. 2. Schema of CATIA

MAS design patterns allow systems to exhibit capabilities that allow agents to act independently, rationally reason and interact within their environment. The BDI framework facilitates two major battlespace functions: detection and recognition and; sensing, perceiving and comprehension of tactical events. CATIA ensures agents exhibit goal-directed behaviours, reasoning and cognitive automation.

4.2 Components of CATIA

CATIA is a Java-based real-time system that utilises and extends BDI4Jade, an open-source Java based BDI framework that extends the Java Agent Development Environment (JADE) [17]. CATIA consists of three main components: *Recognition*, *Reasoning* and *Deliberation*. CATIA also extends the BDI4Jade framework by introducing the concepts of explicit and implicit beliefs, weighted transient desires and intentions. The components of CATIA are discussed in subsequent sections.

Recognition

The *Recognition* component serves CATIA's perception of tactical events and situations. Information is extracted from a tactical event to determine whether: (i) a new tactical situation has occurred, i.e. a new threat or (ii) a new event for the current situation, i.e. new behaviours for a current threat. Perception of a tactical event is illustrated by revising, and updating the current beliefset of a BDI agent. The beliefset of an agent within CATIA consists of implicit and explicit beliefs. Explicit beliefs are derived from the definitive and informative knowledge contained within a tactical event and will invoke the method `getBeliefRevisionStrategy()`. The belief revision will determine whether TIA needs to occur.

The introduction of implicit beliefs in CATIA provides a superior and novel cognitive capability to TIA. An implicit belief is a non-representational environmental attribute or characteristic of an agent that can assist in reasoning, perception and deliberation of a new tactical event. When processing an tactical event, if the informative knowledge does not match the implicit beliefs of the environment or BDI Agent, then a new 'situation' has occurred. The new situation could be the introduction of a new type of threat that is not consistent with the battlespace environment or a new behaviour to a known asymmetrical or smart threat. This introduction of implicit beliefs delivers a superior approach to situation assessment as it is the primary step to reducing uncertainty. The tactical event is delivering information regarding the behaviours of new, or existing threats. Within CATIA, a new belief becomes a `TransientBelief` type and will commence the *Reasoning* cycle within CATIA.

Reasoning

CATIA's *Reasoning* component has two major processes: Comprehension and Representation. Comprehension refers to how the beliefs formulated in the Recognition component are interpreted, integrated and understood. Comprehension requires three stages: Knowledge Retrieval and Selection, Assessment and Representation Update. The first stage aims to utilise battlespace knowledge bases to interpret, integrate and potentially update the information from a BDI Agent's beliefs. This stage can be viewed as memory and learning, as CATIA is relying on past events to formulate an understanding the current situation.

Assessment is the core of Comprehension and can be achieved by using current and novel TIA methodologies. Assessment also utilises sophisticated techniques for contextual knowledge retrieval, which include tactical-to-enterprise technologies (modelling and simulation, or data as a service (MaaS/DaaS)) and, intelligent and dynamic mission data. The result of Assessment is an integrated COP of the current tactical information. In other words, CATIA agents will have a mental model of the current situation.

The next step is the Representation, which contains three processes: Context, Motivation/Goal Creation and Goal Selection. Context is where Transient Goals are initiated. These goals are weighted, where the weights are calculated from pre-existing and dynamic rules. If a transient goal's weighting is below a certain threshold, this could indicate further comprehension is required, thus instantiating a new Comprehension cycle. Goal selection utilises the `OptionGenerationFunction`, and will select a goal based on its weighting. The goal will be created based on the situation's beliefs, meaning several beliefs can be associated with a goal. The selection of the goal will also represent a BDI Agent's desire, or motivational state.

Deliberation

Deliberation is the last component within CATIA and consists of three processes: Goal Feasibility, Plan Creation and Plan Selection. Goal Feasibility will assess whether a goal selected from Reasoning can be achieved. If the goal cannot be achieved, the *Reasoning* component will be instantiated. In this case, the weighting of the goal will be recalculated, another goal will be selected, or the belief may be deemed unacceptable, or no longer required and a belief revision is required. If a goal is feasible, Plan Creation will create `Intentions`. Plan Selection will begin and formulate a Plan strategy that consists of actions. This is where CATIA's decision-making cycle resides.

The decision-making component is complex; and is currently being designed and developed. At the basic level, the decision-making process will determine whether a plan strategy is accepted or rejected. If rejected, Plan Creation and Plan Selection will recommence, thus also instantiating the Reasoning component. The current design of CATIA's decision-making centres around exploiting tactical decision aids within the tactical system to assist in superior decision-making. The decision-making component of CATIA will be discussed in a future paper.

5 CATIA and the Future Integrated Mission System (FIMS)

Current tactical systems are not designed to allow for separate and distinct tactical components, where subsystems have high interdependencies. The result of current tactical system design principles would render a cognitive architecture within current tactical systems near impossible. The Tactical System Integration (TSI) Branch within the Defence Science and Technology (DST) Group has been designing and developing FIMS, a new generation mission system for future air platforms. FIMS is based upon Layered Approach to Service Architectures for a Generic Network Environment (LASAGNE), which is a Service-Oriented Architecture (SOA) framework that delivers a foundation for developers to create concurrent (multi-threaded), interoperable, portable, deterministic and real-time software components. The main advantage of LASAGNE is the ability for dynamic deployment of tactical system components.

A significant advantage of FIMS is its ability to host CATIA, implementing a far more sophisticated, and superior TIA functionality. Currently, if a fusion subsystem cannot determine the threat, it will be classified as unknown. No further assessment or information processing can occur to determine the threat's behaviour or identification in real-time. Rather, manual processing of the threat information is conducted after a mission operation has concluded. CATIA is of great benefit to a fusion component, where upon the identification of an object, or the behaviour of a threat is not persistent with its known behavioural patterns, FIMS would utilise CATIA to further process tactical events related to the threat. A major advantage of this functionality is threat behaviours, identity and intent can be determined in real time, rather than after a mission. This will strengthen the decision superiority of onboard tactical decision aids, but most importantly, enhance the performance of the human operating the tactical system.

6 Future Work and Conclusion

This paper highlights the need for agent-based cognitive architectures and their importance, especially in the area of TIA. In addition, this paper illustrates the current challenges of automated TIA techniques and how SM and MAS can be used to introduce and enhance computational cognitive attributes in situation and threat assessment.

The implementation of cognitive architectures within tactical systems is fast becoming reality. This paper highlights how CATIA is leading a new and emerging area of tactical cognitive architectures. The potential benefits of this exciting and innovative research are extensive, especially for enhancing information and decision superiority within a joint battlespace. CATIA's current development activities are focussed on the *Recognition* and *Reasoning* components. Once completed, they will be integrated within FIMS to investigate the concept of distributed TIA.

The development activities will also shift toward decision-making processes, with an emphasis on distributed tactical and automated decision-making. CATIA will be further developed with a novel approach to decision-making, where the aspiration is to produce a new generation of tactical decision aids to enhance distributed decision superiority.

References

1. Anderson, J.: The Architecture of Cognition. Harvard University Press, Cambridge (1983)
2. Anderson, J., Lebiere, C.: The Atomic Components of Thought. Erlbaum, Mahwah (1998)
3. Blasch, E., Kadar, I., Salerno, J., Kokar, M., Das, S., Powell, G., Corkill, D., Ruspini, E.: Issues and challenges in situation assessment (level 2 fusion). J. Adv. Inf. Fusion **1**(2), 122–139 (2006)
4. Buford, J., Jakobson, G., Lewis, L.: Extending BDI multi-agent systems with situation management. In: 9th International Conference on Information Fusion (2006)
5. Fischer, Y., Bauer, A., Beyerer, J.: A conceptual framework for automatic situation assessment. In: CogSIMA 2011: 2011 IEEE International Multi-Disciplinary Conference on Cognitive Methods in Situation Awareness and Decision Support (2011)
6. Gluck, K.: Cogitive architectures for human factors in aviation. In: Human Factors in Aviation, pp. 375–398. Academic Press (2010)
7. Holsopple, J., Sudit, M.: Enhancing situation awareness via automated situation assessment. IEEE Commun. Mag. **48**, 146–152 (2010)

8. Jakobson, G.: On conceptualization of eventualities in situation management. In: CogSIMA 2013: IEEE International Conference on Cognitive Methods in Situation Awareness and Decision Support (2013)
9. Jakobson, G., Buford, J., Lewis, L.: A framework of cognitive situation modeling and recognition. In: MILCOM 2006: IEEE Military Communications Conference (2006)
10. Jones, R., Wray III, R., van Lent, M.: Practical evalution of integrated cognitive systems. Adv. Cogn. Syst. **1**, 83–92 (2012)
11. Laird, J., Wray III, R.: Cognitive architecture requirements for achieveing AGI. In: AGI-2010: 3rd International Conference on Artificial General Intelligence (2010)
12. Langley, P., Choi, D.: A unified cognitive architecture for physical agents. In: National Conference on Artificial Intelligence (2006)
13. Langley, P., Laird, J., Rogers, S.: Cognitive architectures: research issues and challenges. Cogn. Syst. Res. **10**, 141–160 (2009)
14. Laudy, C., Mattioli, J., Museux, N.: Cognitive situation awarness for information superiority. In: Information Fusion for Command Support (2006)
15. Looney, C., Liang, L.: Cognitive situation and threat assessment of ground battlespace. Inf. Fusion **4**, 297–308 (2003)
16. Newell, A.: Unified Theories of Cognition. Harvard University Press, Cambridge (1990)
17. Nunes, I., de Lucena, J., Luck, M.: BDI4Jade: a BDI layer on top of Jade. In: ProMAS 2011: 9th International Workshop on Programming Multi-agent Systems (2011)
18. Powell, G.: Tactical situation assessment challenges and implications for computational support. In: 8th International Conference on Information Fusion (2005)
19. Salerno, J., Yang, S., Kadar, I., Sudit, M., Tadda, G.P., Holsopple, J.: Issues and challenges in higher level fusion: threat/impact assessment and intent modeling. In: 13th Conference on Information Fusion (2010)
20. Steinberg, A., Bowman, C., White, F.: Rethinking the JDL data fusion levels. In: 2004 National Symposium on Sensor and Data Fusion (2004)
21. Sun, R.: Desiderata for cognitive architectures. Philos. Psychol. **17**(3), 342–373 (2004)
22. Sun, R.: The CLARION cognitive architecture: extending cognitive modeling to social simulation. In: Sun, R. (ed.) Cognition and Multi-agent Interaction: From Cognitive Modeling to Social Simulation. Cambridge University Press (2006)
23. Sun, R.: The challenges of building computational cognitive architectures. Stud. Comput. Intell. **63**, 37–60 (2007)
24. Taatgen, N., Anderson, J.: The past, present and future of cognitive architectures. Top. Cogn. Sci. **2**, 693–704 (2010)
25. Taatgen, N., Lebiere, C., Anderson, J.: Modeling paradigms in ACT-R. In: Cognition and Multi-agent Interaction: From Cognitive Modeling to Social Simulation. Cambridge University Press (2006)
26. Wray, R., Jones, R.: Considering soar as an agent architecture. In: Sun, R. (ed.) Cognition and Multi-agent Interaction: From Cognitive Modeling to Social Simulation. Cambridge University Press (2006)

Security and Trust on Mobile Agent Platforms: A Survey

Donies Samet[1(✉)], Farah Barika Ktata[2], and Khaled Ghedira[3]

[1] Ecole Nationale des Sciences de l'Informatique, Manouba, Tunisia
dadou_2222@hotmail.fr

[2] Institut Supérieur des Sciences Appliquées et de Technologie de Sousse, Sousse, Tunisia

[3] Institut Supérieur de gestion de Tunis, Le Bardo, Tunisia

Abstract. Mobile agent technologies are known for their capacity to develop and construct distributed, heterogeneous and interoperable systems. Despite the presence of several platforms for the development of mobile agent applications, security issues act as a main deterrent against such trends. Based on this, we conducted a comparative study of the most promising and existing mobile agent platforms, showing the diverse security features employed to address various threats. We investigate also the trust models used by the platforms. The established study focuses on the contributions to verify security criteria by the used security mechanisms in every studied platform. This study is important not only to allow practitioners pick the most suitable platforms to meet their security requirements, but also to allow researchers address the voids by ameliorating the concluded limits and proposing possible improvements of new versions of these platforms.

Keywords: Mobile agents · Mobile agent platforms · Security criteria · Trust

1 Introduction

Mobile agent technology offers several advantages for software development such as the reduction of network traffic, disconnected operations and reducing communication costs. Although the mobility of agents has several advantages, it increases security issues. Thus, a key challenge for the expansion of their use is to overcome these problems. To date, in most cases surveys on security in mobile agent systems [1, 2] describe only some basic security characteristics used in some platforms. Moreover in most cases surveys do not provide a comparison of security aspects in mobile agent platforms according to the security criteria and do not take consideration of variants and add-ons. Our paper does not just list the security mechanisms provided by the platforms, but it focuses on security criteria, specific to mobile agent systems, that need to be fulfilled to have a high security level. The criteria, derived from the ISO standard (ISO/IEC 27000:2009), have been detailed to go in depth in the comparison. We have established the correspondence between the defined sub-criteria and the existing mechanisms in each platform. Such an analysis enabled us to identify the location of the gaps in each security system, which made it possible to identify the problem and to

G. Jezic et al. (eds.), *Agent and Multi-Agent Systems: Technology and Applications*, Smart Innovation, Systems and Technologies 74,
DOI 10.1007/978-3-319-59394-4_5

mitigate this fuzzy aspect around security and to identify what is missing to contribute to its insurance. In the analysis tables, each of the boxes that specify that the sub criterion is not verified, present a domain to explore, knowing that the proposal of the right solution begins with the proper specification of the gaps. In this paper the sections are organized as follows: we describe in Sect. 2 the security criteria on mobile agent systems, in Sect. 3 we present a comparative study of mobile agent platforms by investigating the provided security mechanisms and the treated security criteria. In Sect. 4 we present the trust models used by the studied platforms. Finally, the final section presents the conclusion.

2 The Security Criteria on Mobile Agent Platforms

A platform security performance is identified by its level of obedience to the security criteria. Based on [3, 4], we describe the different security criteria on mobile agent platforms as follows:

- Confidentiality: All private data stored on a platform or carried by an agent must remain accessible only to authorized entities.
- Integrity: The platform and the agents must protect from unauthorized changes.
- Access Control: This is the definition of authorizations that specify who is allowed to do what, how and under what condition on an agent and on a platform.
- Availability: The platform of the agent must be able to ensure the availability and usability of data and services to local and remote agents.
- Non-repudiation: The technique of non-repudiation is to eliminate the risk that an agent or a platform may deny sending the data or perform an action.
- Authentication: This is the ability to verify the identity of the two interacting entities that they are what they claim to be.
- Authenticity: The identity verification and at the same time the authentication of the communicating parties and integrity of the transmitted message or agent.

3 Comparative Study of Mobile Agent Platforms

In this section we develop a comparative study of mobile agent platforms with an emphasis on the contribution of provided security mechanisms and approaches to the verification of security criteria. The studied platforms are Aglets [5], SOMA [6, 7], Jade [8, 9] and Cougaar [10, 11]. Even though some platforms do not have recent references, it sometimes presents important solutions and mechanisms that are not used in platforms, which have a higher level of security, with recent references. These platforms are chosen because they are well known and they focus on the aspects of security. In the analysis tables, each of the boxes that specify that the sub criterion is not verified, present a domain to explore, knowing that the proposal of the right solution begins with the proper specification of the gaps.

3.1 The Authentication Criteria

In the current system of Aglets, all servers in the same domain share a secret key, and an agent is authenticated only if it belongs to the same source domain of the visited place. The agent's manufacturer is currently anonymous because code signing is not supported. An extension of aglet SAglet [12] implements Public/Private key for authentication of both the originator and visitor platforms. In SOMA platform, only the agents that their domain of origin is not trusted need to be authenticated. The agent authentication is based on the owner information. Before the agent migration this information, the initial state of the agent and its code are digitally signed by the owner. SOMA uses the role certification standard X.509. Poggi et al. [13] present a security model for JADE that uses a certification authority (CA) and inserts the agent platform into an existing Public Key Infrastructure (PKI). The Jade-S security add-on [8], provides authentication of a user by the verification of its username and password. An extension of Jade-S [14] defines a strong user authentication module using database users' fingerprints that are signed with public/private key pair. The Jade-S has been extended [15] by implementing Login Module and an enhanced Security Certification Authority. The X.509v3 digitally-signed certificates, contains owner biometric information. Another extension of Jade-S [16] adds the self-contained sensors to extend the performance of the authentication system. A PKI add-on [17] is proposed to provide the Jade platform with the possibility of using the digital signature of the home container for the authentication. Cougaar platform allows mutual authentication. An agent can be signed by its owner using X.509 digital certificates or the standard username-password. Cougaar allows the use of public key infrastructure (PKI) and certificating authorities (CA). A summary of the study of the authentication parameters in the mobile agent platforms is presented in Table 1. The platforms which use authentication based on the verification of the agent digital signature allow the receiver to know that only the sender could have encrypted it. Authentication across the platforms is related to the domain (the case of Aglet), the user or the administrator of the place but not the agent. An authentication of the agent itself can allow the consideration of many important security concepts and the definition of fine-grained security policy for a better security level. The fact that a platform does not use a key distribution adds considerable administrative burden and make the system more susceptible to several attacks such as masquerade.

3.2 The Integrity Criteria

In mobile agent systems, the integrity of an agent includes the integrity of the agent code and the agent state (intermediate results). Agent code integrity: All platforms that use digital signature ensure the integrity of the agent code. The digital signature is not only a method of identifying the owner of the agent but it is a guarantee of the integrity of the agent code. The integrity of collected data: Ensuring the integrity of the data is to prevent any entity from changing intermediate results. Another important aspect is the computation integrity: Since a malicious host can execute an agent, it is necessary to have the guarantee that the computation was done according to the instructions of the mobile agent. In this context we make a study concerning the integrity parameters used

Table 1. The authentication parameters in the mobile agent platforms

Platforms		Authentication based on	Key distribution
Aglet	Aglet	Domain secret key	Not available
	SAglet	Digital signature (public/private key of the platform)	Not available
SOMA		Digital signature of the owner	X.509 certification
Jade	Jade-S	Owner username and password (JAAS)	Not available
	[14]	Owner username and password (JAAS), Fingerprints signed with public/private key pair	Not available
	[15]	Owner username and password, digital signature of the agent owner, fingerprint information	X.509 v3 certification
	[16]	Owner username and password, digital signature of the agent owner, fingerprint information, self-contained sensors	X.509 v3 certification
	Jade-S, PKI add-on	Digital signature of the home container	PKI
Cougaar		Username-password, digital signature and X.509 digital certificate of the owner or the server, role based certification	X.509 certification

in the mobile agent platforms (Table 2). Given that the code signing is not provided by Aglet, integrity is not supported directly by Aglets platform. Any malicious agent or platform can change the data an agent carried along the way. SAglet proposes the detection of changes in the agent itinerary using the signature of itinerary object by the originator private key. Secure Aglet Server (SAS) [18] extends Aglets to support the creation of Message Digests using the Java Cryptography Extension (JCE) and to allow the ability to digitally sign objects. SAS adds a java class library that implements the concept of Read-Only data. SOMA presents two solutions to help resolve the problem of the agent state integrity using the Message Integrity Code (MIC). The first solution is that (Trusted Third Party) agents, that need to make cryptographic functions to protect collectable data, must visit a trusted third party place to validate its encryption after visiting an untrusted place. The second solution is that (Multiple-Hops) collected data integrity is assured without imposing any additional visits to other places. Collected data security is ensured by proofs chain structure among the various platforms hosting the agent. The integrity of the "chain" of cryptographic proofs is verified by the owner when the agent returns back. The limit of Multiple-Hops protocol is that it works properly only with the 'visit-once' assumption. Multiple-Hops do not prevent the malicious platform from totally deleting/replacing all data in a chain. SOMA employs the transport layer security (TLS) for the agent transfer. TLS protocol is the successor version of SSL version 3 and is considerably a more secure and capable protocol. In Jade platform, PKI-Add On achieves integrity of the agent code by signing it with the

Table 2. The integrity parameters in the mobile agent platforms

Platforms		Agent code integrity	Collected data integrity	Computation integrity
Aglet	Aglet	Not available	Not available	Not available
	SAglet	Integrity of the agent code by the digital signature, Integrity of the itinerary object (static)	Not available	Not available
SOMA		Integrity of the agent code by the digital signature	Cryptographic functions (TTP, MH)	Not available
Jade	Jade-S	SSL for the agent transfer but agent is not signed	Not available	Not available
	[14]	SSL for the agent transfer but agent is not signed	Not available	Not available
	[15]	Integrity of the agent code by the digital signature, SSL	Not available	Not available
	Jade-S & PKI add-on	Integrity of the agent code by the digital signature of the home container, SSL	Signature of data which is not going to be changed after its initialization	Not available
Cougaar		Agents are shared between platform containers (shared-object mobility)	Agent's state is synchronized across platforms during execution	Real time monitoring of the process execution

private key of the container and attaching the certificate to the field of the agent. Compared to other platforms, Cougaar does not use the process of serialization mobility but it uses shared-object mobility, in which agents are shared between platform containers and the agent's state is synchronized across platforms during execution. The messaging component is used for synchronizing the state: Agents store their state on the Cougaar Blackboard [10]. Cougaar uses assessor and data plugins for real time monitoring of the execution process. The problem of code integrity and collected data, which is not going to be changed after its initialization, is resolved by cryptographic functions and digital signature. The challenge problem to solve is to prevent malicious platforms from making undetectable falsification of the process execution, like pretending using the already calculated intermediate results in the computation of results provided by the malicious platform. Or especially in this case, due to the uncertainty in the computation, we can say that using a trust model, to help decision making process in choosing hosts to visit, can improve considerably the computation integrity by trying to limit the selected hosts to trusted ones.

3.3 Communication Authenticity

Authenticity of a communication between two partners imposes their authentication and the integrity of the transmitted message. Aglet authenticates the communicating parties that belong to the same domain using the secret key but it does not use the exchanged key to encrypt communications. Whiteboard is also a type of communication used in aglets. The SAS aglet extension is selected to implement SSL in the Tahiti server. SOMA uses implicit communication that does not require the name of the communicating parties. Agents in the same place interact by means of shared objects, such as blackboards and tuple spaces [19]. Tuples spaces are suitable for agent coordination but not applicable to bulk data exchange. To provide channel integrity, confidentiality and authentication of communicated parties, SOMA uses TLS protocol with a set of cryptographic algorithm. On Jade platform messages are neither encrypted nor signed by default. It is the responsibility of the agent sending a message to explicitly request the platform to sign and/or encrypt a message. Messages are transferred over secure channels SSL. Cougaar uses Blackboard publish/subscribe for communication within the agent itself and between agents. Inter-agent communication is via message passing. UltraLog components [20] can adaptively support and alternate multiple protocols. Cougaar offers third-party developers the possibility to write new link protocols and plugin them into Cougaar. For all platforms, if a communication is established between two agents when one of them or both are not in their home place, the agents or their home place administrator are not responsible for the authentication of the communicating parties, it is the responsibility of the visited place administrator, or at least the person who specified the certificate used for visited place authentication. The issue here (for the sender and/or the receiver) is about the reliability of these parties and how much an agent can trust a platform to sign or encrypt a message. Different agents from different owners and creators can communicate through the same authenticated secure socket. However, in critical cases this can be considered as an inconvenience. For platforms that use Blackboard it seems there is no existing control related to the content and transaction security is considered only as adherence to objects in the Blackboard.

3.4 Access Control Criteria

In this sub-section we compare the different used access control approaches. There are three hierarchies for Access control in Aglet: general level for unauthenticated manufacturer, organization level for unauthenticated owner and per-aglet level otherwise. Also in Aglet the definition of authorization can use other parameters like computing power, occupancy level, organizational affiliation, pricing, code certification, or the type of aglet. In SAglet, after the verification of the authentication of the visitor agent, a service agent handles the resource request according to the policy file which includes a policy entry to each originator node and agent. The factors taken into consideration when controlling access to resources in SOMA can be divided into two types: static and dynamic factors. Static factors are the identity of the agent and its role. As dynamic factors, we can mention the application state and the resource state. In Jade all actions that agents can perform on the platform are controlled based on the owner identity and according to a policy file. There are two types of policy files: platform and containers policy files. The access control to resources of the platform proposed in [16] uses a merger of three access control methods: Role Based Access Control [21], Mandatory Access Control [22] and Credential Based Access Control [22]. The work of [16] offers the possibility of declaring a role as delegable and/or re-delegable by a delegation certificate to transfer the authorization related to that role to another subject. The access control of agents to resources uses reputation system. Owner reputation values are calculated using a distributed version of the Eigen Trust algorithm [23], based on the agent past behavior and integrity but it is not specific to mobile agent systems. Cougaar uses the KAoS policy system [24]. KAoS allows the definition of high-level policies and finely grained policies. Application-specific concepts can be introduced to the KAOS policy. Policies may be changed by users or by its adaptive security engine. We can conclude that there is no platform which offers an access control policy that treats all parameters in Table 3 and even the parameters used can be improved by taking into account more aspects and details related to mobile agent systems especially parameters related to the trust level.

Table 3. The access control parameters in the mobile agent platforms

Platforms		AgID	OwID	Origin host	Agent Role	Org Aff/gpe	Behavior	Trust level				
								Agent	Owner	Origin host	Org A/G	Itine-rary
Aglet	Aglet	●				●						
	SAglet	●		●		●						
SOMA		●			●							
Jade	Jade-S		●									
	[20]	●	●		●	●	●	●	●		●	
Cougaar		●	●	●	●	●						

Legend:
●: The parameter is used for the definition of the access control policy on the platform.
AgID: Agent ID, OW: Owner ID, Origin Host: Originated host, Org A/G: organizational affiliation/group.

3.5 The Availability Criteria

In Aglet an agent can share a common allowance with its clones and created agent. Allowance gives the possibility of defining time CPU, memory, agent life span, size of an agent and whether new aglets can be created. All these parameters contribute in avoiding blocking. SOMA allows the monitoring of CPU usage percentage; related conditions can be expressed per single target host or per a whole target domain. SOMA can limit the number of operations in case of surcharge. SOMA uses protection domain to isolate agents from other agents running concurrently. In case of intrusion detection, SOMA allows the decrease of the execution priority, the suspension or the termination of the breaker agents. Jade allows the definition of a maximum waiting time for a requested message with the possibility for every single behavior to block its computation, in the meantime, not to waste CPU time. Jade gives the possibility of defining timeout after which an agent must execute a task. Cougaar allows the reactive blocking of ports based on the connection rate. According to the established study (Table 4), there are several aspects that can be improved in mobile agent platforms to increase the level of availability.

3.6 Non-repudiation Criteria

To avoid repudiation a platform should provide a logging service to be able to keep track of all actions and attempts of agents, as well as authentication failures or when transmission faults occur. Aglet uses log files to logs events. SOMA maintains history of previous performance and provides indicators, notifications and statistics of monitored events. Jade provides a logging service by instantiation of logger objects, with the possibility of specifying the logging level and handlers. Logging and analyzing services are among future directions of Cougaar platform.

3.7 Confidentiality Criteria

For agents and messages, the confidentiality is ensured like for any data or component in the network, by the use of cryptographic mechanisms. Aglet proposes to use cryptographic algorithms if the peer contexts want to exchange sensitive information. In SOMA it is possible to use DES channel encryption or the SSLv3, both allows the encryption of agents and messages. Jade platform provide the possibility to encrypt messages using public or private key of the owner. A Jade user just needs to request a message to be encrypted using the EncryptionService. The use of the SSL protocol allows Jade to ensure the confidentiality of the agent during its transfer. Cougaar provides for the developer the possibility to integrate standard practice security measures like cryptographic mechanisms.

4 Trust Models Used in the Platforms

In the Aglets trust model, a server determines that an agent is trustworthy only if it is sent from a server in the same domain. A trusted agent is an authenticated agent. Aglet safety is ensured by restricting the agents transfer to servers in the same domain. The security level is static, an agent is trusted or not for all its lifecycle. In Aglet the visited platform is trusted by default. In SOMA, the notion of trust is equivalent to the concept of role assignment; roles are used to express the trust model chosen by the determination of allowed operations in the resource system. The visited platform is trusted by default. In Cougaar, an authenticated agent is considered as trusted agent. The platform is considered by default trusted by the agents. In the Jade platform an authenticated user is considered to be trusted also his actions and agents that it generates. The platform is considered to be trusted by default. The reputation system described by [16] combines a reputation value to each service provided by each agent which reflects the satisfaction of all the agents they interact with the service. This work use Eigentrust algorithm to calculate the service reputation value. This algorithm is not specific to mobile agent system. In the works of [25, 26], an extension of Jade, the reputation is a

Table 4. Treated aspects to ameliorate availability

Platforms		Control the resource usage	Control related to the agent characteristics	Recovering the system
Aglet		Time CPU/memory usage	Agent's life span, Agent size, whether new aglets can be created	-
SOMA		CPU usage percentage, Limiting the number of operations, Decrease of the execution priority, Suspension or the termination of the breaker agents, Protection domain	-	-
Jade	Jade-S	Avoid active wait	Maximum waiting time for a requested message, timeout for an agent task	-
Cougaar		Reactive blocking of ports	-	-

value earned by a good behavior on a platform. If an agent terminates successfully a session +1 is attributed to his reputation score, −1 otherwise. In Table 5 we present the different aspects of trust management handled by the different platforms. In all studied platforms the hosts are always considered as trusted. This can cause that the threat of a malicious platform be neglected. Even if the platform is authenticated, there is always the possibility that it is or becomes malicious. Also for the agent, the fact that it is authenticated or has a specific role that does not mean it is trusted because it can carry and use malicious code or/and false data. It may have changed its behavior during the course of its itinerary, by an attack carried out by another agent or by a malicious host.

Table 5. The different aspects of trust management handled by mobile agent platforms

Platforms		Static trust	Dynamic trust	Trust computation
Aglet		●		
SOMA		●		
Jade	Jade & Jade-S	●		
	[16]		●	Eigentrust algorithm
	[25, 26]	●		Successful session +1, −1 otherwise
Cougaar		●		

5 Conclusion

In general, all tested platforms use considerable security functions but provide different levels of security. According to this comparative study of platforms, Cougaar and Jade could be considered as platforms that offer higher levels of security than others.

It should be noted that Jade is a platform in accordance with the FIPA standard whereas this is not the case for Cougaar. Jade is more popular and easier to use than Cougaar that provide more complicated and unfamiliar interfaces. The highlight of the Cougaar platform is its Adaptive Security Engine. Cougaar includes a Framework for monitoring and interaction that collects and analyzes data from various entities to detect possible attacks, which can dynamically adjust the security policy in each server. Cougaar is suitable for dynamic and complex environments. The attention to authentication within Cougaar and Jade platforms is very useful for trust management. By this study we conclude that the main problems in mobile agent platforms are: the code or parts of the agent which must be readable by the visited platforms, the encryption and decryption of the messages when the agent is not in its home platform, the case when the collected data will be updated later in the itinerary of the mobile agent before it returns to its home platform and the case when the behavior of an agent depends on results of its execution in foreign hosts. By observing these problems we can conclude that they are all caused by the uncertainty, and providing only 'hard' security is not enough to avoid interacting with an entity that carries malicious code or has a selfish behavior. These observations allow us to say that one of the promoting fields that can bring future amelioration to the next generation of mobile agent platforms, is the trust management as such contribution will allow dealing with uncertainty and avoiding interaction with malicious entity (platform or agent). Integrating a comprehensive trust model specific to mobile agent systems that provides trust decisions in the security decision-making process contribute to enhance and improve security performance. We propose to integrate a framework for trust management based on a trust model specific to mobile agents systems. This framework must include dimensions which allow us to calculate trust level of different entities. A trust level can be calculated based on different metrics like direct interactions, reputation and risk measure. By the elaborated study we conclude that a combination of the most effective mechanisms used by the different platforms with a detailed trust model specific to mobile agent systems and new mechanisms for verification of the sub-criteria identified as unavailable may lead to a new solution with an important level of security. Such a study can also help many potential researchers or developers who want to use mobile agent-based technologies for their applications. With this study, we expect to contribute to cover the need of an updated study related to security level provided by mobile agent platforms, and to encourage future work in the field.

References

1. Aggarwal, S., Bhardwaj, S., Kumar, P.: Security approaches for mobile multi-agent system. Int. J. Emerg. Technol. Adv. Eng. **2**(12), 681–687 (2012)
2. Kravari, K., Bassiliades, N.: A survey of agent platforms. J. Artif. Soc. Soc. Simul. **18**, 11 (2015)
3. Jansen, W., Karygiannis, T.: Mobile agent security, National Institute of Standards and Technology (2000)

4. Braubach, L., Jander, K., Pokahr, A.: A practical security infrastructure for open multi-agent systems. In: 11th German Conference, MATES 2013, Koblenz (2013)
5. Lange, D.B., Mitsuru, O.: Programming and Deploying Java Mobile Agents Aglets. Addison-Wesley Longman Publishing, Amsterdam (1998)
6. Bellavista, P., Corradi, A., Stefanelli, C.: A secure and open mobile agent programming environment. In: Fourth International Symposium on Autonomous Decentralized Systems (1999)
7. Corradi, A., Montanari, R., Stefanelli, C.: Mobile agents integrity in e-commerce applications. In: 19th IEEE International Conference on Distributed Computing Systems (1999)
8. Jade Security Guide (2005). http://jade.cselt.it/doc/tutorials/JADE_Security.pdf
9. Bellifemine, F., Cairea, G., Poggib, A., Rimassac, G.: JADE: a software framework for developing multi-agent applications. Lessons learned. J. Inf. Softw Technol. **50**, 10–21 (2008)
10. Helsinger, A., Thome, M., Wright, T.: Cougaar: a scalable, distributed multi-agent architecture. In: IEEE International Conference on Systems, Man and Cybernetics (2004)
11. Feiertag, R., Rho J., Rosset S.: Using security mechanisms in Cougaar, 1st Open Cougaar Conference (2004)
12. Nusrat, E., Ahmed, A.S., Rahman, G.M., Jamal, L.: SAGLET-secure agent communication model. In: 11th International Conference on Computer and Information Technology, ICCIT (2008)
13. Poggi, A., Rimassa, G., Tomaiuol, M.: Multi-user and security support for multi-agent systems. In: WOA (2001)
14. Conti, V., Vitabile, S., Pilato, G., Sorbello, F.: An enhanced authentication system for the JADE-S platform. WSEAS Trans. Inf. Sci. Appl. **1**, 178–183 (2004)
15. Vitabile, S., Pilato, G., Conti, V., Gioè, G., Sorbello, F.: Biometric features for mobile agents ownership. IPSI BgD Trans. Internet Res. **1**, 81–89 (2004)
16. Vitabile, S., Conti, V., Militello, C., Sorbello, F.: An extended JADE-S based framework for developing secure multi-agent systems. Comput. Stan. Interfaces **31**, 913–930 (2009)
17. http://jade.tilab.com/doc/tutorials/PKI_Guide.pdf
18. Jean, E., Jiao, Y., Hurson, A.R., Potok, T.E.: SAS: a secure aglet server. In: Computer Security Conference (2007)
19. Cabri, G., Leonardi, L., Zambonelli, F.: Auction-based agent negotiation via programmable tuple spaces. In: 4th International Workshop, CIA, Boston, MA, USA (2000)
20. http://ultralog.net
21. Sandhu, R.S., Coyne, E.J., Feinstein, H.L., Youman, C.E.: Role-based access control models. Computer **29**, 38–47 (1996)
22. Sandhu, R.S., Samarati, P.: Access control: principle and practice. IEEE Commun. Mag. **9**, 40–48 (1994)
23. Kamvar, S.D., Schlosser, M.T., Garcia-Molina, H.: The Eigentrust algorithm for reputation management in P2P networks. In: Proceedings of the 12th International Conference on World Wide Web (2003)
24. http://www.ihmc.us/research/projects/KAoS/
25. Singh, A., Ahuja, P.: Robust algorithm for securing an agent hosting platform. Int. J. Adv. Technol. **3**, 84–91 (2012)
26. Ahuja, P., Sharma, V.: A JADE implemented mobile agent based host platform security. Comput. Eng. Intell. Syst. **3**, 8–19 (2012)

A Self-adaptive System for Improving Autonomy and Public Spaces Accessibility for Elderly

Sameh Triki[1(✉)] and Chihab Hanachi[2]

[1] 118 Route de Narbonne, 31062 Toulouse, France
sameh.triki02@gmail.com
[2] 2 Rue du Doyen-Gabriel-Marty, 31042 Toulouse, France
hanachi@univ-tlse1.fr

Abstract. Nowadays, there is an increasing need to provide a safe and independent living for cognitively deficient population. Notably, we have to improve seniors' autonomy and their public spaces accessibility. Giving these observations, the aim of this paper is to provide a personalized adaptive assisting system for elderly. More precisely, this paper presents the specification and implementation of a self-organizing multi-agent system able to abstract the different distributed components involved in user's environment. This system is able to detect different possible situations that a user could face in his daily outdoors activities and propose accordingly appropriate actions. This system not only learns user's habits from its perceptions but also improves its recommendations thanks to feedbacks provided by stakeholders (family, doctors ...) following a reinforcement learning reasoning. Finally, we present our system evaluation specially its learning capabilities through different scenarios that have been generated automatically.

Keywords: Assisted living system · Multi-agent system · AMAS theory · Reinforcement learning

1 Introduction

Currently, the increasing ageing of the population is one of today's major problems. Some difficulties, such as cognitive deficiency among seniors, make the independent access to the city difficult and unfortunately encourage them to stay at home [1, 2]. The increasing population of elder people and their social isolation requires that more activities should be done in order to improve their quality of life. One possible way is to provide them with tools that assist them.

Nowadays, the rapid increasing of electronic components and the reduction of their cost have led to an explosion of the number and functionalities of smart devices. Applications of these devices have reached various domains and a considerable amount of progress was noticed in assisting seniors in their life such as home monitoring, fall detection and geolocation gadgets. However, the majority of devices on the market, are designed for indoor care or limited to a defined zone or made to assist a limited set of predefined tasks [3–5]. Moreover, they do not provide a personalized tool that is able to

G. Jezic et al. (eds.), *Agent and Multi-Agent Systems: Technology and Applications*, Smart Innovation, Systems and Technologies 74,
DOI 10.1007/978-3-319-59394-4_6

evolve according to the user needs and to adapt to their gradually cognitive decline or sudden habits changes [6].

This paper presents a research that aims at designing and developing an adaptive accompanying system that enables potentially vulnerable and dependent population to maintain their social life and to ease their access to urban services independently and to unsure user security while being outside. The targeted population is older adults having age-related cognitive deficiencies such as memory loss, difficulties in doing parallel tasks and activities planning.

We propose a self-organizing multi-agent system called Sadikikoi[1] able to perceive its environment and detect conventional or problematic situations. Our system will propose actions to respond to detected situations and will improve its behavior to adapt to the user habits and requirements by a reinforcement learning approach. The system should be able: to evolve using received perceptions coming from different components, to analyze and correlate those perceptions and to propose the action required for assisting the user in his daily life activities outdoors. Moreover, perceptions may be coming from a smartphone sensor, wearable sensors or other devices distributed in the environment. Thus a multi-agent system seems adequate to handle those different components and their interactions and to deal with scalability. Our multi-agent system combines the AMAS ("Adaptive Multi-agent System") theory and a reinforcement learning approach.

Our contribution is threefold. First, we provide a multi-agent architecture: its components and their interactions described with a UML diagram. Second, we define the self-adaptive feature of our system that is based on self-organization following a reinforcement approach. The self-organization feature relies on the automatic creation, modification or deletion of agents without external intervention. We give high-level algorithms and a UML sequence diagram to specify this feature. Third, we provide and discuss evaluations based on simulations.

The paper is organized as follows. Section 2 describes the state of the art of existing researches sharing a similar goal such as assisted living systems and their limitation. Section 3 shows the proposed multi-agent system, its architecture and the specification of its self-adaptive feature. Section 4 illustrates the implemented system and its usefulness. Section 5 concludes the paper.

2 Related Work

Several advanced researches on ambient assisted living (AAL) systems have already been done. In this section, we will compare the most representative AAL systems of the state of the art and specifically the closest ones (same target population, close functions …) to our research.

We consider eight systems (AMON, WEALTHY, UCS, COACH, PEAT, Autominder, OutCare and KopAL) that are compared considering two disciplines (see Table 1): computer science and social human science. Regarding social science,

---

[1] The system build is called Sadikikoi and it is part of the compagnon project funded by the Midi-Pyrénées Region.

Table 1. Classification of AAL systems

	Disciplines	Criteria	Sub-criteria	Ambient Assisted Living systems (AALs)					
				AMON [4]	WEALTHY [7]	UCS [8]	coach [5]	peat/autominder [9/10]	OutCare/kopAL [11,12]
SHS	USAGER	functional requirements (provided services)	autonomie	-	-	+	+	+	+
			mobility	--	--	++	--	--	+
			protection	--	--	+	--	--	+
		user needs and unfunctionnal requirements	vulerability	+	+	+	+	-	+
			acceptance	+	+	+	+	+	-
			transparency	-	++	-	--	-	-
	ENVIRONNEMENT	interaction with environmental components		--	-	+	-	--	-
		context aware		-	-	-	-	--	+
		covered scope (complete/limited zone)		--	-	+	--	+	--
INFORMATIQUE	METHODOLOGY	interdisciplinarity		-	-	+	-	-	-
		participative design (user centred approach)		+	-	+	-	-	+
	SYSTEM REQUIREMENTS	*adaptation/evolution*		--	--	--	--	--	--
		heteregenity		-	-	-	-	-	-
		ambient		-	-	++	-	--	+

we consider user criteria including functional requirements and user needs. Regarding computer science, we take into account system requirements and the followed methodology (interdisciplinary, participative design).

AMON a wearable multi-parameter medical monitoring and alert system [4] and WEALTHY a Wearable Health Care System Based on Knitted Integrated Sensors [7] consist on wearable sensors that provide continuous monitoring of their user. Both share the same functionality: save user's information and send a signal. The information transmitted is stored in a server and then sent either to family, a neighbor or a helping person. Those devices are certainly interesting, though, is not adaptive but rather reactive device: when a defined condition is detected, a signal is sent. So this system will neither learn nor evolve according to the changing needs of its user.

Japan's Fujitsu UCS has unveiled a prototype cane equipped with a GPS that guides the seniors in their movements, but also to monitor them remotely and monitor their heart rate [8]. As this rod is still in the prototype stage there is not much details about it neither about the used techniques. It is definitely interesting but it relies on sending information and receiving control guidance based on predefined or communicated thresholds.

Many other assisting devices are defined for indoors assistance, some are usually targeting a specified task such as the COACH system (Cognitive Orthosis for assisting aCtivities in the Home) [5] that guide a user through hand washing task based on planning, other cognitive orthotics tools such as PEAT [9] and Autominder [10] also use automated planning to provide generic reminders about daily activities.

Others targets a limited zone: OutCare [11] and KopAL [12] support outdoor wandering issues related to disorientation by alerting the caregiver when leaving a predefined routes or deviating from daily usual routes.

Table 1 classifies the different described ambient assisted living devices AALs according to defined criteria extracted according to the user needs and requirements to create a useful and usable assisting device.

As described in Table 1[2], the majority of AAL systems are designed to respond to predefined situations with predefined conditions and circumstances but they are not targeting unpredictable situations. Even though they are targeting the same population and aiming at assisting elderly, those works differ from our work since they are not adaptive.

We discussed the use of those devices in healthcare based on individuals' medical conditions, such as physical or mental disabilities, chronic disease, or rehabilitation situations. One of the most important shortcomings is that those related works are not adaptive, nor personalized and tailored to the needs of each user.

Thus, it becomes essential to have an evolving system that will adapt to the user change of requirements.

This is not an easy task given that aside social issues and ethics of such a device, the goals set up are not easily reachable. It is very difficult to test exhaustively a device capable of responding to unpredictable situations.

3 A Self-adaptive Multi-agent System

In order to design and build a self-adaptive system to assist seniors in their daily outdoor activities while respecting their requirement, we have involved users and their surroundings, human social science (HSS) and medical experts during the design and development process following a participative approach.

3.1 Theoretical Foundation of Our Proposed Approach

The computing environment where our system should be deployed is complex since it is *distributed*, includes a large number of entities and constraints (devices, sensors …) and *mobile* users and devices. It is *open* since components can enter and leave the system at any time.

To deal with distribution and openness constraints, we use a multi-agent approach. As for unpredictability and adaptation, our agents will follow a reinforcement learning approach taking into account the context.

Let us precise the interest of context and learning in our approach before detailing the proposed approach.

Need of context

Our system should adapt to the changing contexts of the user and take into account his possible evolving requirements. In other words, we need to build a context-aware software that could self-adapt according to user habits, his dependency level, the changing environment and accessible devices. *The notion of Context* has been widely studied (see for example [14, 15]). In our case, we define the context not only by the user's location but also the correlation of several parameters such as: network

---

[2] Table legend: The "+ or ++" symbol shows how much the feature is taken into account by each system and shortcomings of each system are spotted with "– or ––" symbol.

connectivity, noise level, heartbeats rate, and every possible sequence of our system perceptions (events, environment …). These parameters could be collected through smartphone sensors or connected wearable devices. As stated in [15], the context provides a meaning to a situation and help to choice the more relevant action to trigger. In our case, the context is the sequence of our system perceptions (events, environment …) that justifies the birth of adequate actions.

Need of learning
Giving the numerous and possibly unpredictable contexts that could exist in our complex environment, it is difficult to enumerate them a priori. One way to deal with this issue is to use a machine learning approach that enables to build systems that automatically improve their functioning with experiences and learn to adapt to new possible contexts. Machine learning is classified in three categories [16, 17] and in our work, we will use the same reasoning as in the reinforcement learning approach. It is inspired from the natural learning that learns from its actions and mistakes to produce better performance in the future, through rewards. Also, it is the only approach that doesn't require examples and deals with dynamicity and the unpredictability, required for our system. More precisely, our agents will automatically determine the ideal behavior for a specific context based on feedbacks from the environment, and they will keep on adapting their behavior through time. The goal of each agent is to maximize its total reward while being cooperative. We will detail the learning process and their interaction protocol in Sect. 3.2.2.

Proposed Multi-agent Approach: AMAS
We use the Adaptive Multi-Agent Systems (AMAS) [13] approach that aims at solving problems in a dynamic non-linear environment by enabling agents to learn their cooperative behavior in order to make emerge the global function of the system. AMAS theory has presented encouraging results in several context aware application such as user monitoring system indoors [18] and boat behavior detection [19].

In AMAS theory, agents self-organize by cooperation since automatic creations, modifications, merging or deletions of agents are operated without external interventions. Applying this approach to the problem of context learning leads to a specific type of agents, called Context Agents. They are created at runtime and self-adapt on-the fly [20].

To be cooperative means to valorize the global goal over the personal goal. For example, if a context agent becomes useless, it will self-destroy for the benefit of the system. Agents face Non-Cooperative Situation (NCS). For example, a NCS is detected when two context agents propose two different actions in the same situation and the wrong recommendation was selected. In this case, a resolution to solve the NCS situation is conducted collectively by the agents of the system.

3.2 Elderly Monitoring and Assistance with AMAS: System Architecture and Self-adaptive Feature

3.2.1 Architecture

We implemented a multi-agent system that offers monitoring and analysis of deviation to detect disorientation situation and assist the user through notifications, alerts and to identify and recommend good practices. This system also keeps a margin of initiative to the user and allows the emergence of innovative processes (set of actions).

Our system contains three main types of agents which are: the percept Agent, The context Agent and the Effector Agent types. Figure 1 describes the architecture of our system and its interactions with the environment.

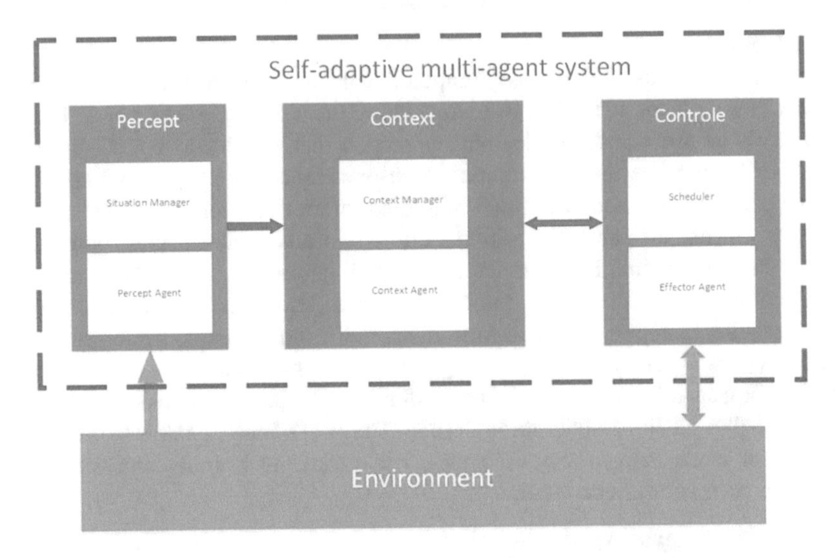

Fig. 1. The different modules of our system and its environment

We have three modules in our system that communicates and collaborates, each composed of a manager and a set of agents. At first, the system contains one situation manager, one context manager and one scheduler. Percept, context and effector agents are created progressively at run time.

The percept module detects, saves and interprets inputs received directly from the environment. Moreover, it guarantees the routing of significant signals to the context module.

The Context module contains context agents that reason on sent perceptions and propose appropriate actions.

The Control module gathers the scheduler (decision maker) and effector agents (performers). It receives context agents' action propositions, select the most appropriate one and execute it. It acts and receives feedbacks from the environment for learning purpose. By environment, we mean both the user behavior and the physical environment (park, ways, restaurants …) and stakeholders.

Testing our system with real users is risky and time consuming at this stage because they are vulnerable, not always available and can't behave normally when observed. So we decided to define a virtual environment perceived through distributed and unpredictable simulated events. These events are organized by scenario. Each scenario represents a succession of user daily possible outdoors activities and system actions that have a relation: done during a given period of time, made on the same place.

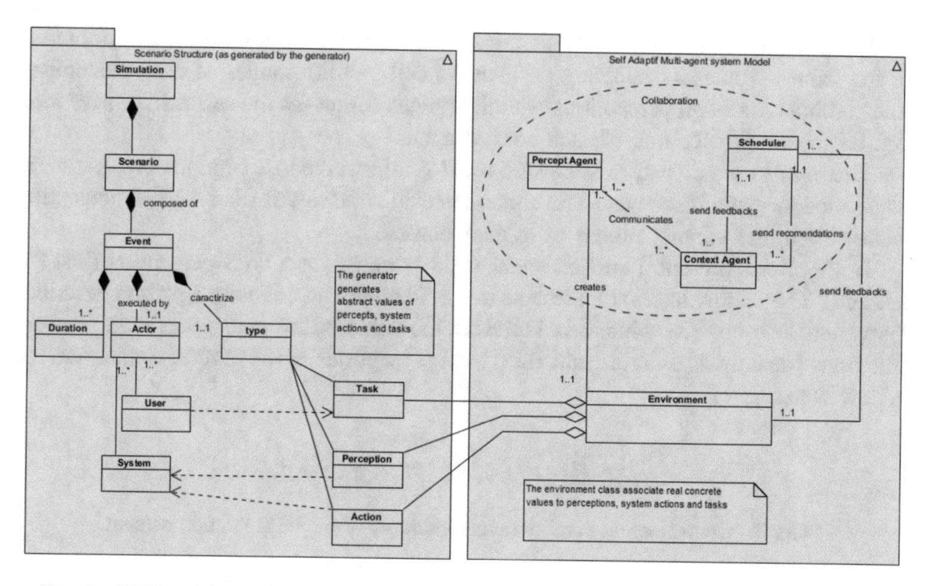

Fig. 2. Different components of the virtual environment and our system (UML diagram)

Figure 2 presents the different components of our MAS (right hand side of the figure) and the composition of the environment. This later corresponds to the structure of the scenario (Left hand side of the figure).

In the left hand side of Fig. 2, we represent the structure of generated scenarios composing our virtual environment. A scenario is a sequence of events (e.g. tasks, perceptions and actions) made by an actor (e.g. user and system) for a defined duration. On the right hand side, we can clearly see interactions between the three types of agents (e.g. percept agents, context agents, effector agents).

3.2.2 The Self-adaptive Feature of Our System

In this section, we will specify the context agent and his cooperative behavior that represents the key component for the self-adaptive feature of our system.

A. *The context Agent specification*

Firstly, each context agent has four attributes: *context*, *state*, *action*, *appreciation*.

Structure of a Context Agent
Context: *{<perception1, valueInf, valueSup>, ..., <perceptionN, valueinf, valuesup>}*
State: *(created/validable/valid/selected/dead)*
Action: *action to be executed*
Appreciation: *confidence rate*
End Structure

Let us detail its attributes: Context agent associates to each perceptions an interval of the values [valueInf, valueSup], which we call validity ranges of each perception. The combination of all perceptions' validity ranges composes the *context. Appreciation* defines the confidence rate of each context agent.

The context agent state will be said ***valid*** if all received perceptions' values are in its defined the validity ranges. The context is said ***validable*** if the received perception values are in the validity ranges or in their borders.

In Fig. 3 we present a context agent validity range for a GPS perception (just the latitude). The yellow interval represents the valid zone, the one with a pattern describes the validable zone. The white box represent the value of the perception in the current situation. The context agent is said valid if all perceptions of the current situation in the yellow zone.

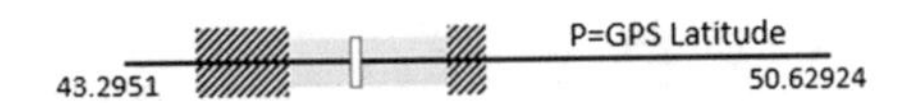

Fig. 3. Context agent validity ranges example of the GPS latitude percept

Each context agent follows a life-cycle, at first it is created. Then it can propose an action if it is in a validable or a valid state. It can be selected by the scheduler and have its action executed. It can be in the dead state if its confidence is too low. The system faces four *Non-Cooperative Situations (NCS)*: ***NCS1***: System unproductivity (No context agent proposing the desired action), ***NCS2:*** A wrong action has been performed, ***NCS3***: Useless context Agent has been detected, ***NCS4***: Conflict between context agents.

To better understand those NCS, the sequence diagram in Fig. 4 describes the resolution of NCS1.

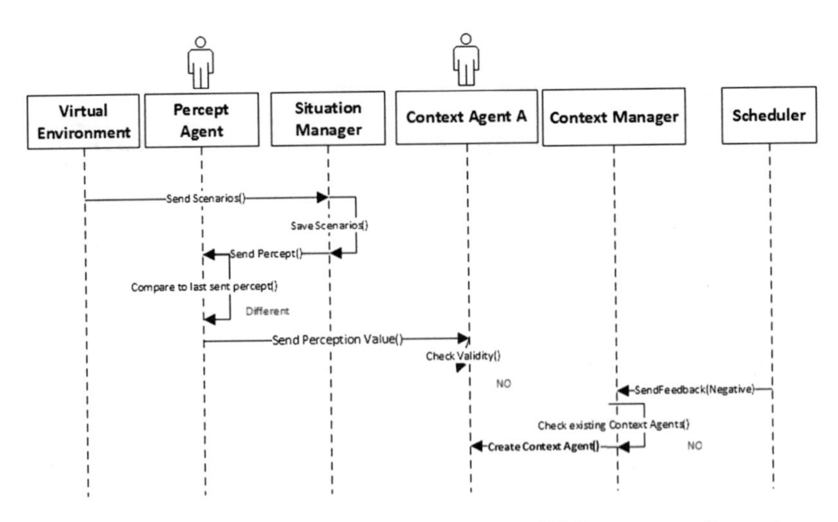

Fig. 4. System unproductivity resolving process (UML sequence diagram)

NCS1 will be managed by the context manager based on the scheduler feedbacks and will create a new context agent that provides the desired action for an interval that contains the perceptions at that time. We will detail behaviour algorithms and how it manages the NCS.

B. *Behavior algorithms ensuring the self-adaptive feature*

The first algorithm describes the scheduler. It have two phases, the first on is to select the recommendation to be executed and then to send feedbacks according to the performed action and detects different types of NCS. However, it's not responsible of neither their occurrence nor their resolution and it just follows a couple of rules when facing multiples propositions:

- The most confident context agent is selected and its proposition is executed
- The valid context is automatically chosen over a validable context

So context agents should manage to make the context agent proposing the relevant action be valid and the most confident.

Algorithm 1. Scheduler Behaviour Algorithm

```
Require: Sch Scheduler, Propositions are received from each context agent valid or validable
Require: C_s Selected Context Agent, A_s Proposed action of the selected Context Agent C_s
Require: A_d desired Action, C_d Context Agent proposing A_d
 1: while Sch.ExistsPropositions() do        ▷ Step1: Recommendation Selection and execution
 2:     if Sch.ExistsValidContext() then
 3:         C_s=Sch.selectMostConfident(validContexts);
 4:     else
 5:         C_s=Sch.selectMostConfident(allContexts);
 6:     end if
 7:     Sch.SendActionToExecute(A_s);
 8: end while
 9: while Sch.ExistsFeedbacks() do            ▷ Step2: Recommendation Evaluation
10:     if (existsCorrectContext()) then
11:
12:         if ( A_s == A_d ) then             ▷ Case of good recommendation
13:             Sch.sendFeedback(C_s, Positive);
14:         else                                ▷ Case of failing recommendation
15:             Sch.sendFeedback(C_s, Negative,C_d);
16:         end if
17:     else
18:         if Sch.ExistsSelectedContext() then
19:             Sch.sendFeedback(C_s, Negative);
20:         else
21:             Sch.sendFeedback(Negative);
22:         end if
23:     end if
24: end while
```

There are different three of negative feedbacks, one when there is a context agent that proposes the desired action along received propositions "*sendFeedback*(C_s, *Negative*, C_d)",one where none of the proposed context agents proposes the desired action "*sendFeedback*(C_s,*Negative*)" and one that no context was proposed when an action was needed "*sendFeedback*(*Negative*)". In the two latter cases, the context manager will be involved in the resolution of those NCSs. The second algorithm explains the behavior of the context manager whose main role is to dispatch the received feedbacks to the corresponding context agents. Moreover, the context manager resoles NCS1 and NCS2 by creating a new context agent which explains how our system becomes populated with context agents in the first place.

Algorithm 2. Context Manager Behaviour Algorithm

```
Require: CM Context Manager,C_i Context Agent i, Feedbacks are received from the scheduler
1: while CM.ExistsFeedbacks() do
2:     CM.DispatchFeedbackToContextAgent();
3:     if ReceivesFeedback(C_i,Negatif) then
4:         CM.CreateNewContextAgent(Confidence_i); /*Confidence_newContext > Confidence_i*/
5:     end if
6:     if ReceivesFeedback(Negatif) then
7:         CM.CreateNewContextAgent();
8:     end if
9: end while
```

The latter algorithm describes the nominal behavior of the context agent (perceptions receptions, state update and recommendations) and the cooperative behavior when repairing NCS. The context agent ultimate goal is to always be in a cooperative situation.

Algorithm 3. Context Agent Behaviour Algorithm

```
Require: C_i Context Agent i, C_n Context Agent n in conflict with C_i,
Require: Perceptions received from each percept agent,Feedbacks received from the scheduler
1: while true do                      ▷ Step1: Recommendation according to received perceptions
2:   C_i.ReceivesPerceptions();
3:   C_i.UpdateState()
4:   if (C_i.isValid() or C_i.isValidable()) then
5:     C_i.proposeRecommendation (Action_i;Confidence_i;State_i)
6:   end if
7:   while C_i.ExistsReceivedFeedbacks() do          ▷ Step2: Learning from received feedbacks
8:     if (C_i.ReceivesFeedback(C_i,Positive)) then  ▷ Dealing with positive feedback in case of good recommendation in a previous step
9:       if ( C_i.isValid()) then
10:        C_i.IncreaseConfidence ();
11:      else
12:        if C_i.isValidable then
13:          C_i.IncreaseValidityRange ();
14:        end if
15:      end if          ▷ Dealing with negative feedback in case of failing recommendation in a previous step
16:    else                                 ▷ Case of failing recommendation from a valid context
17:      if ( C_i.isValid()) then
18:        C_i.decreaseValidityRange ();
19:        C_i.UpdateState() ;
20:
21:        if C_i.isValid then
22:          C_i.decreaseConfidence ();
23:        end if
24:        if (C_i.checkConfidence () <=0) then
25:          C_i.Self-destruct (); /*NCS3 resolving*/
26:          EXIT;
27:        end if
28:      end if     ▷ Case of failing recommendation while an other context agent proposed the desired action
29:      if (C_i.ReceivesFeedback(C_i,Negatif,C_n)) /*NCS4 resolving*/ then
30:
31:        if (C_i.isValid() AND C_n.isValid()) then
32:          C_i.SendIncreaseConfidence (C_n,Confidence_i); /*Confidence_n > Confidence_i*/
33:        end if
34:        if (C_i.isValid() AND C_n.isValidable()) then
35:          C_i.SendIncreaseValidityRange(C_n);
36:          C_i.SendIncreaseConfidence (C_n,Confidence_i); /*Confidence_n > Confidence_i*/
37:        end if
38:        if (C_i.isValidable() AND C_n.isValidable()) then
39:          C_i.SendIncreaseValidityRangetoValid (C_n);/*Until C_n becomes valid*/
40:        end if
41:      end if
42:    end if
43:  end while                                   ▷ Step3: Learning from received messages
44:  while C_i.ExistsReceivedMessages() do
45:    if (ReceivesIncreaseValidityRange()) then
46:      C_i.IncreaseValidityRange ();
47:    end if
48:    if (ReceivesIncreaseValidityRangetoValid()) then
49:      while ( C_i.isValidable()) do
50:        C_i.IncreaseValidityRange ();
51:      end while
52:    end if
53:    if (ReceivesIncreaseConfidence()) then
54:      C_i.IncreaseConfidence ();
55:    end if
56:  end while
57: end while
```

We have two main hypothesis to ensure cooperative behavior between context agents while dealing with NCSs which are: a ***validable*** context agent does not update its confidence, but updates its validity ranges and a ***valid*** context agent can update both its confidence and its validity ranges.

The context agent behavior algorithm consists on an infinite loop made of three steps. First, it will start its nominal behavior by updating his state according to received perceptions and possibly propose a recommendation (line 2 to 6). Second, the context agent starts a loop to treat received feedbacks from the scheduler in order to learn the user's habit and adapt to his requirement. There are two types of possible feedbacks, positive corresponding to correct recommendation leading the concerned agent to increase its confidence or its validity ranges (line 8 to 15). Negative feedbacks are reactions to failing recommendations which trigger NCS (line 16 to 42). Third, the context agent cooperates with other context agent by executing sent suggestions (line 44 to 56). Feedbacks being sent by the scheduler, the second algorithm details the behavior of the scheduler.

The self-adaptation feature is based on cooperation that reduces NCS and therefore maximizes good recommendations by means of self-organization, consisting in:

1. ***Creating*** new context agents in case of system unproductivity
2. ***Modifying*** existing ones through self-adjustments on the confidence and validity range
3. ***Reducing*** the number of context agents by self-destruction of useless context agent and by merging context agents when they propose the same action for the same situation.

Now that we have an overview on our system process, in the following section we will describe our application and different primary results.

4 System Evaluation

As explained in Sect. 3, our system was tested through scenarios generated automatically. The generator simulates a user's activities through a combination of perception, action and tasks. Tasks correspond to user's activities such as "going shopping", "walking", doctor's visit.

Figure 5 illustrates the configuration panel of our generator in which we define various parameters (i.e. number of generated days, number of actions and perceptions in scenario …) of the generator corresponding to user's characteristics (his autonomy level, system acceptance level, …).

We generated a set of 100000 scenarios and created a simulation of 1000 days for a single user profile. We observe the behavior of our system to test self-adaptation and self-organization by checking when the system will be able to propose the action required. The generator goal is to create enough scenarios for a given period of time (e.g. a day, a week …) in order to evaluate the learning capabilities of our system.

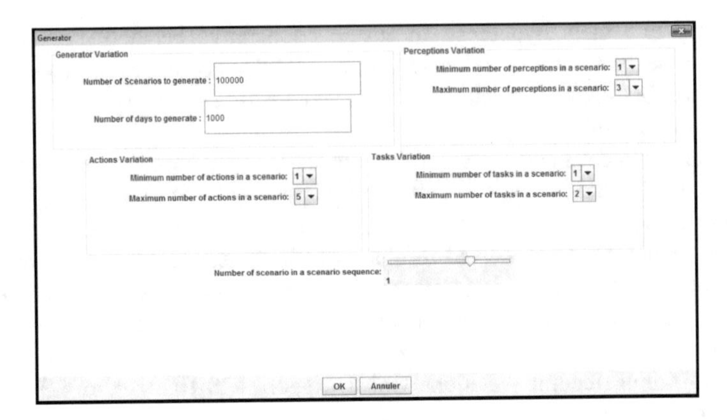

Fig. 5. Configuration panel of the scenarios generator

Figure 6 describes the system considering four common behaviors of learning process:

- **Right True:** Is when the system proposes the right action
- **Wrong True:** Is when the system proposes an action that should not have been proposed
- **Right False:** Is when the system doesn't proposes any actions and it was the right
- **Wrong False:** Is when the system doesn't proposes any actions but it should have.

These three graphs of Fig. 6 are generated automatically throughout the execution of our system. The horizontal axes represent the number of steps (corresponding to the number of received events). We can clearly see that all graphs start at zero since no recommendation can be made when no context agent is created yet. The first graph shows how context agent number evolves with events. We can notice that at the end we

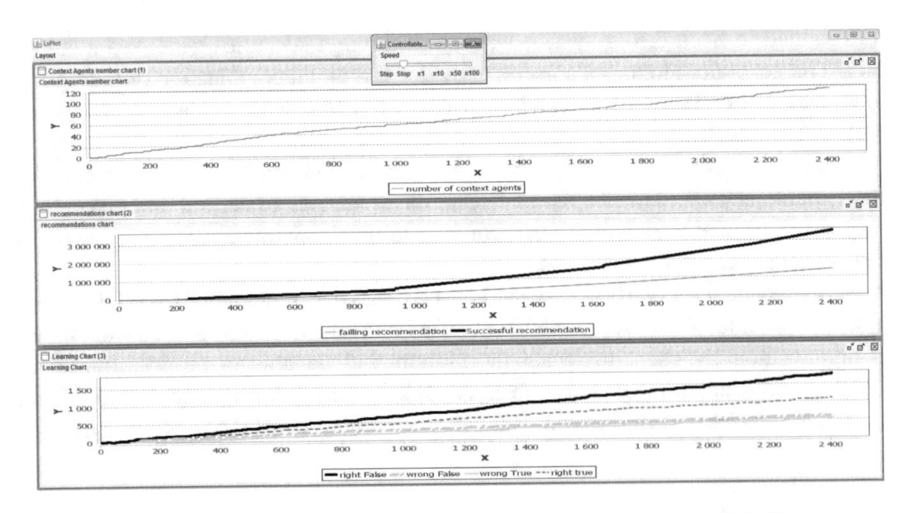

Fig. 6. Variation of right and wrong predictions made by the MAS

have 120 context agents for 2400 events. These numbers show that our system doesn't create an agent for each event but a same agent is able to manage several correlated events (corresponding to a situation) thanks to its adaptation.

In the two last graphs of Fig. 6, we can see that our system performs well: Correct recommendations are almost three times greater than the failing ones. The fact that the number of errors keeps increasing comes from the unpredictably of our daily life scenarios. The last graph shows how the correct recommendations (right false and right true) suggesting the right behavior are three times greater than wrong recommendations (wrong false, wrong true).

5 Conclusion and Perspectives

In this paper, we have shown that current Ambient Assisted Living systems (AALs) are facing strong limitations and lacking adaptation since they do not take into account environment and users' requirements evolutions.

We have defined the architecture of a self-adaptive multi-agent system that can assist older adults in their outdoor activities and handle unexpected situations. Our system has the capacity to learn user habits from its perceptions and improve its recommendations thanks to feedbacks provided by stakeholders. A simulator has been implemented and experimentations have shown the feasibility of our system and the efficiency of our learning process.

However, it is clear that there are still a number of issues that require further investigations. The simulator can be improved to take into account the user planning in its recommendations in order to provide recommendation compliant with the user planning. Moreover, we intend to implement a prototype on smartphones and test it with real users.

Acknowledgements. We would like to acknowledge both the Midi-Pyrénées Region and the IRIT laboratory that funded the Compagnon project. We also would like to thank Marie-Pierre Gleizes from IRIT Laboratory and Alice Rouyer from LISST laboratory for helping in specifying the system developed.

References

1. United Nations: Population Division. World Population Ageing. Department of Economic and Social Affairs (2013)
2. Fondation of France. 2010–2014: The French increasingly only (in French)
3. Rashidi, P., Cook, D.J.: Keeping the resident in the loop: adapting the smart home to the user. IEEE Trans. Syst. Man Cybern. A Syst. Hum. **39**(5), 949–959 (2009)
4. Anliker, U., et al.: AMON: a wearable multiparameter medical monitoring and alert system. IEEE Trans. Inf Technol. Biomed. **8**(4), 415–427 (2004)
5. Mihailidis, A., Carmichael, B., Boger, J.: The use of computer vision in an intelligent environment to support aging-in-place, safety, and independence in the home. IEEE Trans. Inf Technol. Biomed. **8**(3), 238–247 (2004)

6. Sadri, F.: Ambient intelligence: a survey. ACM Comput. Surv. **43**(4), 36:1–36:66 (2011)
7. Paradiso, R., Loriga, G., Taccini, N.: A wearable health care system based on knitted integrated sensors. IEEE Trans. Inf Technol. Biomed. **9**(3), 337–344 (2005)
8. Ninomiya, J., Murayama, K., Yamaoka, T., Sakai, K.: Next-generation ubiquitous device for new mobility society: next generation Cane. FUJITSU Sci. Tech. J. **51**, 8–13 (2015)
9. Levinson, R.: The planning and execution assistant and trainer (PEAT). J. Head Trauma Rehabil. **12**(2), 85–91 (1997)
10. Pollack, M.E., Brown, L., Colbry, D., McCarthy, C.E., Orosz, C., Peintner, B., Ramakrishnan, S., Tsamardinos, I.: Autominder: an intelligent cognitive orthotic system for people with memory impairment. Robot. Autonom. Syst. **44**(3–4), 273–282 (2003)
11. Wan, J., Byrne, C., O'Hare, G.M., O'Grady, M.J.: Orange alerts: Lessons from an outdoor case study. In: Proceedings of 5th International Conference on Pervasive Computing Technology for Healthcare (2011)
12. Fudickar, S., Schnor, B.: Kopal a mobile orientation system for dementia patients. Intell. Interact. Assist. Mob. Multimedia Comput. **53**, 109–118 (2009)
13. Capera, D., Georg´e, J.-P., Gleizes, M.-P., Glize, P.: The AMAS theory for complex problem solving based on self-organizing cooperative agents. In: Enabling Technologies: Infrastructure for Collaborative Enterprises, WET ICE 2003 (2003)
14. Bazire, M., Brézillon, P.: Understanding context before using it. In: Dey, A., Kokinov, B., Leake, D., Turner, R. (eds.) CONTEXT 2005. LNCS, vol. 3554, pp. 29–40. Springer, Heidelberg (2005)
15. Salembier, P., Dugdale, J., Myriam, F.: A descriptive model of contextual activities for the design of domestic situations. In: European Conference on Cognitive Ergonomics (2009)
16. Mitchell, T.M.: Machine Learning. McGraw Hill, New York (1997). ISBN: 0-07-042807-7
17. Mitchell, T.M.: The discipline of machine learning, vol. 9. Carnegie Mellon University, School of computer science, Machine learning department (2006)
18. Guivarch, V., Camps, V., Péninou, A., Glize, P.: Self-adaptation of a learnt behaviour by detecting and by managing user's implicit contradictions. WI-IAT **3**, 24–31 (2014)
19. Brax, N., Andonoff, E., Gleizes, M.-P.: A self-adaptive multi-agent system for abnormal behavior detection in maritime surveillance. In: Jezic, G., Kusek, M., Nguyen, N.-T., Howlett, Robert J., Jain, Lakhmi C. (eds.) KES-AMSTA 2012. LNCS, vol. 7327, pp. 174–185. Springer, Heidelberg (2012)
20. Boes, J., Nigon, J., Verstaevel, N., Gleizes, M.-P., Migeon, F.: The self-adaptive context learning pattern: overview and proposal. In: Christiansen, H., Stojanovic, I., Papadopoulos, George A. (eds.) CONTEXT 2015. LNCS, vol. 9405, pp. 91–104. Springer, Cham (2015)

Meaning Negotiation with Defeasible Logic

Matteo Cristani[1(✉)] and Antonino Rotolo[1,2]

[1] Dipartimento di Informatica, Università di Verona,
Cà Vignal 2, strada Le Grazie 15, 37134 Verona, Italy
matteo.cristani@univr.it
[2] CIRSFID and Law School, University of Bologna, Bologna, Italy
antonino.rotolo@unibo.it

Abstract. Agents negotiate the meaning of terms in numerous real-life situations. When they behave so, they can be used as the basis for providing an emulation paradigm for software agents, habilitating therefore socio-technical systems to perform meaning negotiation. In this paper we focus upon two methods for meaning negotiation in defeasible logic and provide room for an analysis of how the proposed approaches perform the aforementioned process. Finally we also provide a computational analysis of the process automation problem.

1 Introduction

In general, negotiation is a dialogue between two or more agents by which they try to reach an agreement about something starting from different viewpoints about the shared object. A negotiation process is *quantitative* when the agents discuss about how to share a set of countable objects, whereas in *Meaning Negotiation*, on which we focus in this paper, the proposals are pieces of knowledge represented by terms, i.e. the expressions of what an agent knows about the negotiated terms. These pieces of knowledge may be accepted or rejected by the other discussants.

More specifically, Meaning Negotiation (henceforth MN) is a negotiation process in which the sharing object is the meaning of a set of terms. An agent R (the receipient) tries to reach an agreement with a counterpart C by looking at a proposal π made by C that R compares with her viewpoint ω. The comparison produces an evaluation that can produce different effects:

- R can accept π for it is *compatible* with ω. The negotiation ends with an agreement;
- R rejects π for it is *incompatible* with ω, but since R thinks that it is possible to obtain an agreement on a different proposal that she hopes can be derived in the negotiation process, she responds with a new proposal. This action inverts the turns in the negotiation process, so that at next turn the role played by R is played by C and vice versa;

G. Jezic et al. (eds.), *Agent and Multi-Agent Systems: Technology and Applications*, Smart Innovation, Systems and Technologies 74,
DOI 10.1007/978-3-319-59394-4_7

- R rejects C's proposal, and since she came to the conclusion that it is not possible to reach any agreement with C, she terminates the negotiation process with an explicit disagreement with C's viewpoint.

In [5–8] it has been deeply analysed, in different contexts, how to reach agreements, and have been provided methods to implement algorithms that provide negotiation processes automatically. Although this approach has revealed to be quite fruitful, only preliminary results has been obtained so far that exhibit deduction system architecture models for meaning negotiation [9].

In this paper we focus upon a mechanism that is somewhat intrinsic in defeasible logic that naturally mimics meaning negotiation, the distinction between derivation and conclusion, that is also the base of non monotonic capabilities of defeasible logic. In particular we study a notion of proposal and viewpoint as studied before for meaning negotiation and embed them in the machinery of defeasible logic, by means of the notions of literal set and theory. Further on, we study two variants of defeasible logic systems, that can be used to provide room for strategies of negotiation and compatibility. These are, respectively, obtained by introducing revision operators and repair chains. These concepts allow to determine new models of meaning negotiation that are very simple in terms of machinery and also efficient computationally.

The purpose of the paper is therefore to clarify how to employ the internal mechanisms of defeasible logic to emulate the behaviour of a software agent negotiating the meaning of a set of terms. This allows to employ a very powerful tool that is also close to logic programming, in order to define such agents.

The rest of the paper is organised as follows. In Sect. 2 we present the formalism of defeasible logic, and in Sect. 2.1 we accommodate meaning negotiation in defeasible logic. Section 3 takes some conclusions and sketches further work.

2 Defeasible Theories as Negotiation Structures

DL is based on a logic programming-like language and it is a simple, efficient but flexible non-monotonic formalism capable of dealing with many different intuitions of non-monotonic reasoning. DL is closely related to logic programming [2] and an argumentation semantics exists [11]. DL has a linear complexity [18] and also has several efficient implementations (e.g., [3]).

Defeasible logic has originally been created by Donald Nute with a particular concern about efficiency and implementation.

The main intuition of the logic is to be able to derive "plausible" conclusions from partial and sometimes conflicting information. Conclusions are tentative conclusions, in the sense that a conclusion can be withdrawn when we have new pieces of information.

Defeasible Logic is a sceptical formalism, meaning that it does not support contradictory conclusions. In cases where there is some support for concluding ϕ but also support for concluding $\neg\phi$, the reasoning mechanism does not conclude either of them. The contradiction is resolved by a priority relation between the supports. A *defeasible theory*, i.e. a knowledge base in Defeasible Logic, consists of two kinds of knowledge: facts and rules. *Facts* denote simple pieces of information that are deemed to be true regardless of the other knowledge items. They are indisputable statements represented by ground literals. For example, "Tweety is a penguin" is represented by $Penguin(Tweety)$ A *rule* describes the relationship between a set of literals, *premises*, and a literal, *conclusion*. Rules specify the strength of the relation between premises and conclusion by having different kinds of rules. Rules distinguish between *strict rules*, *defeasible rules* and *defeaters* representing respectively, by expressions of the form $A_1, \ldots A_n \rightarrow B$, $A_1, \ldots A_n \Rightarrow B$, $A_1, \ldots A_n \rightsquigarrow B$, where $A_1, \ldots A_n$ is a possibly empty set of prerequisites and B is the conclusion of the rule.

Strict rules are rules in the classical sense: whenever the premises are true then so is the conclusion. They are used for definitional clauses and the conclusion of a strict rule is a fact. An example of a strict rule is "Penguins are birds", formally: $Penguin(X) \rightarrow Bird(X)$;

Defeasible rules are rules that can be defeated by contrary evidence. The meaning of a defeasible rule is "whenever the premises are true then presumably so is the conclusion". An example of defeasible rule is "Birds usually fly": $Bird(X) \Rightarrow Fly(X)$. The idea is that if we know that X is a bird, then we may conclude that X can fly *unless there is other evidence suggesting that she may not fly*;

Defeaters are special kind of rules. They are used to prevent conclusion, not to support them. For example: $Heavy(X) \rightsquigarrow \neg Fly(X)$ states that if something is heavy then it might not fly. This rule prevents the derivation of $Fly(X)$ and it cannot be used to support a $\neg Fly(X)$ conclusion.

The prevention from concluding contradictions is made by the *superiority relation* among rules. The superiority relation is used to define priorities among rules, that is, where one rule may override the conclusion of another rule. For example: $Bird(X) \Rightarrow Fly(X)$ and $r' : Penguin(X) \Rightarrow \neg Fly(X)$ which contradict one another, no conclusive decision can be made about whether a Tweety can fly or not. But if a superiority relation is introduce as $r' \succ r$ then it can be concluded that Tweety cannot fly since it is a penguin.

Conclusions can be classified as definite or defeasible. A definite conclusion is a conclusion that cannot be withdrawn when new information is available. A defeasible conclusion is a tentative conclusion that might be withdrawn by new pieces of information. In addition the logic is able to tell whether a conclusion is or is not provable. Thus it is possible to have the following 4 types of conclusions:

- Positive definite conclusions: meaning that the conclusion is provable using only facts and strict rules;
- Negative definite conclusions: meaning that it is not possible to prove the conclusion using only facts and strict rules;
- Positive defeasible conclusions: meaning that the conclusions can be defeasible proved;
- Negative defeasible conclusions: meaning that one can show that the conclusion is not even defeasibly provable.

Strict derivations are obtained by forward chaining of strict rules, while a defeasible conclusion A can be derived if there is a rule whose conclusion is A, whose prerequisites (antecedent) have either already been proved or given in the case at hand (i.e., facts), and any stronger rule whose conclusion is $\neg A$ (the negation of A) has prerequisites that fail to be derived. In other words, a conclusion A is (defeasibly) derivable when:

- A is a fact; or
- there is an applicable strict or defeasible rule for A, and either
  - all the rules for $\neg A$ are discarded (i.e., not applicable) or
  - every applicable rule for $\neg A$ is weaker than an applicable strict or defeasible rule for A.

Alternatively the reasoning process can be explained in terms of arguments with a three phase process:

1. Give an argument for the conclusion to be proved;
2. Consider all possible counter-arguments for the conclusion;
3. Rebut the counter-arguments:
   - Show that a counter-argument is not valid (e.g., some of the premises do not hold);
   - Defeat a counter-argument by a stronger argument supporting the conclusion.

A defeasible theory $\mathcal{D}$ is a triple $\langle \mathcal{F}, \mathcal{R}, \succ \rangle$ where $\mathcal{F}$ is the set of facts, $\mathcal{R} = \mathcal{R}_s \cup \mathcal{R}_d \cup \mathcal{R}_{dft}$ is the set of strict, defeasible rules and defeaters, and $>$ is the superiority relation on $\mathcal{R}$.

A rule $r \in \mathcal{R}$ is formed by an antecedent or body $A(r)$ and a consequent or head $C(r)$. $A(r)$ consists of a finite sequence of literal while $C(r)$ contains a single literal.

We denote by $\mathcal{C}(q)$ the set of rules having q as consequence.

A conclusion derived form the theory $\mathcal{D}$ is a tagged literal and is categorised accordingly to how the conclusion can be proved:

- $+\Delta q$ if q is *definitely provable* (only facts and strict rules are used in the derivation);
- $-\Delta q$ if q is not *definitely provable* (it is different from proving that $+\Delta\neg q$ so that it is definitely provable that $\neg q$);
- $+\partial q$ if q is *defeasibly provable*, then only defeasible rules and/or defeaters are used in the derivation;
- $-\partial q$ if q is not *defeasibly provable.*

A derivation in defeasible logic is a finite sequence $P = (P(1), P(2), \ldots, P(n))$ of tagged literals. Each tagged literal satisfies some proof conditions. A proof condition corresponds to the inference rules corresponding to one of the four kinds of conclusions we mentioned above. $P(1 \ldots i)$ denotes the initial part of the sequence P of length i. In Table 1 we state the conditions for strictly and defeasible derivable conclusions.

Table 1. Proof conditions for definite and defeasible derivations.

$+\boldsymbol{\Delta}$: If $P(i+1) = +\Delta q$ then $\exists r \in \mathcal{R}_s[q]$ $\forall a \in A(r) : +\Delta a \in P(1 \dots i)$	$-\boldsymbol{\Delta}$: If $P(i+1) = -\Delta q$ then $\forall r \in \mathcal{R}_s[q]$ $\exists a \in A(r) : -\Delta a \in P(1 \dots i)$
$+\partial$: If $P(i+1) = +\partial q$ then 1. $+\Delta q \in P(1 \dots i)$ 2. (a) $\exists r \in \mathcal{R}_{sd}[q] \forall a \in A(r)$: $+\partial a \in P(1 \dots i)$ and (b) $\forall p \in \mathcal{C}(q). -\Delta p \in P(1 \dots i)$ and (c) $\forall s \in \mathcal{R}[\mathcal{C}(q)]$: i. $\exists a \in A(s) : -\partial a \in P(1 \dots i)$ or ii. $\exists b \in \mathcal{R}_{sd}[q]$ such that $\forall a \in A(b) : +\partial a \in P(1 \dots i)$ and $b \prec s$	$-\partial$: If $P(i+1) = -\partial q$ then 1. $-\Delta q \in P(1 \dots i)$ 2. (a) $\forall r \in \mathcal{R}_{sd}[q] \exists a \in A(r)$: $-\partial a \in P(1 \dots i)$ or (b) $\exists p \in \mathcal{C}(q) + \Delta p \in P(1 \dots i)$ or (c) $\exists s \in \mathcal{R}[\mathcal{C}(q)]$: i. $\forall a \in A(s) : +\partial a \in P(1 \dots i)$ and ii. $\forall b \in \mathcal{R}_{sd}[q]$ such that $\exists a \in A(b) : -\partial a \in P(1 \dots i)$ or $b \not\prec s$

For the purposes of this paper we can restrict ourselves to defeasible theories without defeaters. The knowledge representation assumptions we make here can be summarised in the point below. Before to go into that analysis, let us recall more formally what kind of model for negotiation we can assume. Basically, by following the approach proposed in Sect. 1 we aim at allowing two contendants to discuss about a set of literals, with the goal of deriving a subset of non-contradictory literals that result acceptable for both parties under an evaluation semantics that is retained private by each of the two contendants.

Every contendant has a private viewpoint, that is used to evaluate literals. We explore two basic ways of performing this process, that are both derived by defeasible logic research:

- By means of belief revision (in particular base revision [10]) as applied to the superiority relation;
- By means of the superimposition of a logical model in which discussants relate parts of the viewpoint by preferences on what part has to be applied to try the derivation of a literal when the previous part failed in this process.

Within the above described framework, we make the following assumptions:

- A *viewpoint* is a defeasible theory;
- A *proposal* is a finite set of literals;
- The *evaluation* of a proposal is a function that returns *true* when a proposal is derived in a viewpoint, and *false* in the opposite case.

The behaviour of agents is modelled by the operations that need to be performed to manage the process described in [7]. An agent's decision when passing on a proposal is *nondeterninistic* and should be discriminated in a strategic standpoint, that has already been shown to be a difficult task to accomplish [14]. We avoid here to treat this aspect, that is however important and will be investigated in further work.

2.1 Defeasible Reasoning as Negotiation

When an agent receives a proposal, she can evaluate it positively, and this means that his viewpoint derives each of the literals in the proposal. When, instead, it is evaluated negatively, there can be different conditions. Let us consider an agent A and one agent B, that are, in turn, receipient and counterpart in the negotiation process. B makes a proposal π to A who evaluates it negatively.

- In the proposal π there is a literal l that is opposite to a literal in the set of strictly derivable literals of the viewpoint of A. The agent A definitely rejects the proposal. Moreover, the agent also performs a *compatibility analysis* to establish whether it is or not the case to make a counterproposal basing this decision on the history of the proposals $\pi_1, \pi_1, \ldots, \pi_n$ the two parties have made so far;
- In the proposal π there is one literal that is opposite to a literal in the set of defeasibly derivable literals of the viewpoint of A. In this case the agent A shall decide whether it is the case or not to make a counterproposal, basing this decision on the history of the proposals $\pi_1, \pi_1, \ldots, \pi_n$ the two parties have made so far. This may depend upon the *compatibility strategy* that is used by A (that can be different from the one adopted by B);
- In the proposal of π there are only literals that cannot be derived by the viewpoint of A but are not opposite to any literal derived by A. In this case the agent A can accept the proposal, but, based on her *negotiation strategy* she can also decide to make a counterproposal.

The idea underlying the model of negotiation presented above is that an agent can accept a proposal when she can derive it entirely, or when she can *contract* his viewpoint so that the viewpoint derives the proposal. On the other hand we can assume that when a proposal is made, it necessarily contains *all the strictly derivable* literals of the proposer. In this way, when two viewpoints are essentially incompatible, for they contain at least one literal each that is opposite to one literal of the other's viewpoint, no proposal can be accepted and the negotiation ends after one single step.

The strategies to obtain a contraction that results compatbile with other's viewpoint, can be limited. For instance, we can decide that the only possible way to modify a viewpoint is by making a revision of the superiority relation, as in [13]. More generally speaking, we could also extend the strategy to allow adding rules, or substituting, but this would make the operational sequence creative, limiting the intrinsic meaning of the notion of viewpoint. In fact, if the revision process might be creative, consequently, a viewpoint could simply be a set of admissible literals, not a set of literals to which an agent commits. Therefore, either we can contract rules freely, or we can act in a free way, but only on the superiority relation. On the other hand, it is clear that the family of revision processes that we are considering here is not general, as we exclude the revision of facts. In literature, belief revision only involving facts, is usually named *base revision*.

The most general possible way to provide a strategy is to explore nondeterministically the contractions that bring to a theory compatible with the literals. While doing so, we can memorize the resulting defeasible theories and compare them against the set of literals proposed at the negotiation stage. There can be two results:

- The set of literals in the proposal can be accepted only by contractions that have already been used to derive proposals that have been rejected. In this case, the agent rejects the proposal and ends the negotiation;
- The set of literals in the proposal can be accepted by a contraction that have not yet been made. Every proposal satisfying this condition can be used as a counterproposal.

On the above reasoning we base the notion of negotiation strategy. Before to do so, however, we also need to incorporate a notion that has been deeply studied in Deontic Logic and Defeasible Logic literatures: the one of *repair chain*. The idea is to represent the actions taken in a legal system to repair the result of the violation of a law. This may consist of other rules to be applied instead, and can ultimately establish what are the consequences of the violation (for instance in systems with punishment). For a general study of problems related to the repair chain as applied to defeasible logic, in particular with applications to business process compliance see [12,15].

2.2 Base Revision Operators

Several authors have studied base revision and applied to rule systems and defeasible logic as well. In particular in [17] authors have introduced some basic operators to revise rule systems used to represent legal systems. The fundamental operators are named *abrogation* and *annulment*. These operators are derived from the classical contraction operator as defined in monotonic systems in [1], and further developed in [4,16]. We are interested in a simplified version of the contraction operator: the nondeterministic operator of contraction. Consider a proposal π made by the counterpart C and evaluated by the recipient R, by means of her viewpoint T, a defeasible theory $\langle \mathcal{F}, \mathcal{R}, \succ \rangle$. By $E^{+\partial}(T)$ we denote the set of all literals that can be derived defeasibly in T. A nondeterministic contraction operator $T^{contr(l)}$, contracts the theory T by the literal l in such a way that

$$T^{contr(l)} = \begin{cases} T & \text{if } l \in E^{+\partial}(T) \\ T' \in \{(F, R', \prec')\} & \text{otherwise} \end{cases} \tag{1}$$

Where the rule set R' is obtained by contracting (namely, by removing rules) R, and so is the set $\prec'$, derived by contraction from $\prec$. The theory T' is such that, in T', $-\partial l$. Trivially, if in T, $-\partial l$, then $T = T'$. The proof of the following claim is

Proposition 1. *Given a defeasible theory T, there exist finitely many T' that contract T by $T^{contr(l)}$, for every literal l, as much as* $2^{(\|R\| \cdot \| \prec \|)}$.

Another interesting computation is the number of possible minimal contractions. A contraction $T' = \{(F, R', \prec')\}$ is minimal if and only if in every contraction $T'' = \{(F, R'', \prec'')\}$ such that $R' \subseteq R''$ or $\prec' \subseteq \prec''$, $+\partial l$. The maximum number of minimal contractions can be easily computed, as it consists in all the element contraction on R and $\prec$, namely the set of contractions that can be obtained by removing one element in R or one in $\prec$.

Proposition 2. *Given a defeasible theory T, there exist minimal contractions of T by $T^{contr(l)}$, for every literal l, not more than* $(\|R\| + \| \prec \|)$.

Proof. A minimal contraction consists in removing only one rule or one preference pair. Clearly, it may be the case that we obtain not a correct minimal contraction by doing so, therefore the figure is maximal.

It might seem, at first, that this provides room for a linear time solution of the problem of computing a minimal contraction, but this is not the case, as it might happen, instead, that you need to visit, in a complete fashion, the set of possible contractions of a given theory, not only the minimal, to determine one contraction that does not derive anymore a given set of literals. We can prove, however, rather easily, that, with a direct oracle on the combinatorics of elements in R and $\prec$, we can determine a minimal contraction in a polinomial time on nondeterministic machines. Therefore we have the following result, whose proof is omitted, for the sake of space.

Theorem 1. *The problem of computing the minimal contraction of a given set of literals for Defeasible Theories is in NP.*

This may be a rather unpleasant aspect of the developed system. However, there is a natural answer to the arising problem. This consists in limiting the set of contractions that can be used by user's preferences. We might for instance be interested only onto element contraction. Obviously this is a partial negotiation strategy as it can fail in providing all the possible contractions that are candidate to prevent the derivation of one literal. This is coherent with the fact that the complexity of strategic argumentation is NP-complete, as proved in [14], a result that is strictly connected with the notion of contraction of theories. In a certain manner, we might say that strategic argumentation is the dual of meaning negotiation. In strategic argumentation discussants rule themselves within a discussion that can bring to opposite conclusions, and provide arguments that support one thesis and disprove opponent's counterthesis; in meaning negotiation, we aim at agreeing on the same conclusions about the world, and therefore on the meaning of the derived terms, but without agreeing on the ways in which these conclusions are made. Most obviously, if we determine a subset S of theories that we aim at exploring, we can state the following.

Theorem 2. *The problem of computing nondeterministically the theory contraction of a defeasible theory T by S chosen contractions is $O(\|S\|)$.*

Rather obviously, the set S can have superpolynomial size in the number of involved literals. The representation of the above can be made by the construction of *repair chains*, by means of the relation $\otimes$. The expressions we introduce in the logical theory $T = (F, R', \prec', C)$ where C is a set of repair chains $r_1 \otimes r_2 \otimes \ldots r_n$, and $\prec (r_{i_1}, r_{j_1}) \otimes \prec (r_{i_2}, r_{j_2}) \otimes \prec (r_{i_n}, r_{j_n})$ where r_i is an element of the set R, and $\prec (r_{i_x}, r_{j_y})$ means that the relation $\prec$ is established between r_{i_x} and r_{j_y}. The meaning of the repair chain $r_1 \otimes r_2 \otimes r_n$ is that, if the element contraction of r_i shall be explored before the contraction of $r_{(i+1)}$ for every $1 \leq i \leq (n-1)$.

We can easily prove the following combinatorial result, where we assume that the maximal length of a repair chain of rules and preferences is denoted by $m(T)$ for a theory T, while the number of these chains in denoted by $n(T)$.

Theorem 3. *The problem of computing nondeterministically the theory contraction of a defeasible theory T with repair chains is $O(n(t) \cdot m(T))$.*

3 Conclusions

This paper has discussed the notion of meaning negotiation as embedded into defeasible logic. We studied two variants of the logical system that allow revision of knowledge and repair chains. We obtain some basic computational results, in particular regarding the basic process of implementing the negotiation and the compatibility strategies. The results are discouraging from a general viewpoint and foresee room for the provision of intermediate models that result interesting from a practical purpose. In particular we argue that the simplest strategy of negotiation and the simplest strategy of compatibility analysis are combined in a rational manner and show that these are polynomially solvable on deterministic machines.

References

1. Alchourrón, C.E., Gärdenfors, P., Makinson, D.: On the logic of theory change: partial meet contraction and revision functions. J. Symbolic Logic **50**, 510–530 (1985)
2. Antoniou, G., Billington, D., Governatori, G., Maher, M.J.: Embedding defeasible logic into logic programming. Theory Pract. Logic Program. **6**(6), 703–735 (2006)
3. Bassiliades, N., Antoniou, G., Vlahavas, I.: A defeasible logic reasoner for the semantic web. Int. J. Seman. Web Inf. Syst. **2**(1), 1–41 (2006)
4. Boella, G., Pigozzi, G., van der Torre, L.: Agm contraction and revision of rules. J. Logic Lang. Inf. **25**(3–4), 273–297 (2016)
5. Burato, E., Cristani, M.: Contract clause negotiation by game theory, pp. 71–80 (2007)

6. Burato, E., Cristani, M.: Learning as meaning negotiation: a model based on english auction. In: KES-AMSTA 2009 Proceedings of the Third KES International Symposium on Agent and Multi-Agent Systems: Technologies and Applications. LNCS, (LNBI, LNAI), vol. 5559, pp. 60–69 (2009)
7. Burato, E., Cristani, M.: The process of reaching agreement in meaning negotiation. In: Nguyen, N.T. (ed.) Transactions on Computational Collective Intelligence VII. LNCS, vol. 7270, pp. 1–42. Springer, Heidelberg (2012)
8. Burato, E., Cristani, M., Viganò, L.: A deduction system for meaning negotiation. In: Omicini, A., Sardina, S., Vasconcelos, W. (eds.) DALT 2010. LNCS, vol. 6619, pp. 78–95. Springer, Heidelberg (2011)
9. Burato, E., Cristani, M., Viganó, L.: Meaning negotiation as inference. arXiv preprint arXiv:1101.4356 (2011)
10. Giusto, P., Governatori, G.: A new approach to base revision. In: Barahona, P., Alferes, J.J. (eds.) EPIA 1999. LNCS (LNAI), vol. 1695, pp. 327–341. Springer, Heidelberg (1999)
11. Governatori, G., Maher, M.J., Billington, D., Antoniou, G.: Argumentation semantics for defeasible logics. J. Logic Comput. **14**(5), 675–702 (2004)
12. Governatori, G., Olivieri, F., Rotolo, A., Scannapieco, S., Cristani, M.: Picking up the best goal. In: Morgenstern, L., Stefaneas, P., Lévy, F., Wyner, A., Paschke, A. (eds.) RuleML 2013. LNCS, vol. 8035, pp. 99–113. Springer, Heidelberg (2013)
13. Governatori, G., Olivieri, F., Scannapieco, S., Cristani, M.: Superiority based revision of defeasible theories. In: Dean, M., Hall, J., Rotolo, A., Tabet, S. (eds.) RuleML 2010. LNCS, vol. 6403, pp. 104–118. Springer, Heidelberg (2010)
14. Governatori, G., Olivieri, F., Scannapieco, S., Rotolo, A., Cristani, M.: Strategic argumentation is np-complete. Front. Artif. Intell. Appl. **263**, 399–404 (2014)
15. Governatori, G., Olivieri, F., Scannapieco, S., Rotolo, A., Cristani, M.: The rationale behind the concept of goal. Theory Pract. Logic Program. **16**(3), 296–324 (2016)
16. Governatori, G., Rotolo, A.: Changing legal systems: abrogation and annulment Part I: revision of defeasible theories. In: Meyden, R., Torre, L. (eds.) DEON 2008. LNCS, vol. 5076, pp. 3–18. Springer, Heidelberg (2008)
17. Governatori, G., Rotolo, A.: Changing legal systems: Legal abrogations and annulments in defeasible logic. Logic J. IGPL **18**(1), 157–194 (2010)
18. Maher, M.J.: Propositional defeasible logic has linear complexity. Theory Pract. Logic Program. **6**(1), 691–711 (2001)

Artificial Intelligence Techniques for the Puerto Rico Strategy Game

Rafał Dreżewski[1(✉)] and Maciej Klęczar[2]

[1] AGH University of Science and Technology,
Department of Computer Science, Kraków, Poland
drezew@agh.edu.pl

[2] The School of Banking and Management, Faculty of Management,
Finance and Computer Science, Kraków, Poland
maciekkl@o2.pl

Abstract. It was always a challenging task to create artificial opponents for strategy video games. It is usually quite easy to discover and exploit their weaknesses because their tactics usually do not adapt to changing conditions and to human opponent tactics. In this paper two artificial intelligence techniques for well known Puerto Rico strategy game are proposed. One of them does not rely on any precoded tactics, but tries to dynamically learn and adapt to the changing game environment. Both techniques were compared on the basis of results of games played against each other and also against human expert players.

Keywords: Strategy games · Artificial intelligence techniques · Computer games

1 Introduction

Since the invention of computers, Artificial Intelligence (AI) was always a topic of high interest. Research on AI was not only related to computer games—in fact it was rather conducted as a part of academic studies or military development. In the recent decades, along with the rapid growth of video games industry, AI for games started to become a necessary part of games development. Almost every published game has some simple—or more sophisticated—AI algorithms inside, which are responsible for actions performed by various gaming agents.

In games that require rather agility skills (e.g. shooting games, hack and slash games), AI development is focused on creating more human-like agents, that are competitive, but not too difficult to compete with, allowing human players to eventually win. If the task is to calculate some number based on the given variables (e.g. hit the target accurately) AI player will outplay human player quite easily, so in most cases the main difficulty is to properly handicap AI player (make it less perfect), in order to achieve more human-like behavior.

The situation becomes completely different in the case of games that require thinking, inventing new tactics, adapting to environment, properly reacting to

G. Jezic et al. (eds.), *Agent and Multi-Agent Systems: Technology and Applications*, Smart Innovation, Systems and Technologies 74,
DOI 10.1007/978-3-319-59394-4_8

enemy's moves. While the above skills would be appreciated in many types of games, strategy games definitely require these skills. That is why it was always hard to create a good AI technique for strategy games [8]. AI techniques for enterprise video games usually are based on instructions what to do in particular situations. This is usually achieved by finite state algorithms or scripts. Skill of such AI depends on how many possible situations the AI's creator would predict. However, in most strategy games it is impossible to predict all situations or combinations, or even get close to it. Such an approach leads to weak AI technique for games. Usually, AI would behave correctly in some situations and much worse in the others. Even bigger problem is that such AI will often perform the same actions in the same or similar situations and its moves would be easy to predict, while prediction of enemy moves is a key of every successful tactic. That is why a good player will sooner or later find such weaknesses and exploit them in order to easily win the game.

The goal of this paper is to present a different approach—AI player that can quickly adapt and change its tactics dynamically on the basis of changes of the game environment. This is achieved by using tree search technique, with such adjustments, that would minimize search time, while keeping results at the satisfying level. Such AI player is capable of competing with an advanced human player and of making its decisions within a reasonable time slot, which is very important for a good game experience. In order to test different proposed AI techniques the computer game based on Puerto Rico board game was implemented. Also two AI algorithms for that game were proposed and implemented in order to illustrate the difference between more common approach (based on precoded decisions) and the proposed adaptive approach.

2 Related Work

Due to the complexity of the most of strategy games, creating AI for them was always very difficult—that is why AI techniques for most enterprise games are quite simple, usually based on scripts. Such an AI will not be a serious challenger for any advanced player, therefore the common technique of improving its skills is providing it a number of advantages (more money, more units, less penalties, etc.). However such an approach is not ideal—basically it is cheating.

Of course, everyone can point to chess game, where AI have achievements of beating the best human players. But there are some significant differences between chess game and enterprise strategy games. Chess AI has been a subject of university studies for almost 70 years. There are a lot of strategies available and that strategies can be hard-coded into chess AI algorithms. The lower bound of the game-tree complexity of chess is 10^{120}, which makes it possible to search all states for next few moves in a reasonable time, or even prepare a database with some precalculated moves. In most of the enterprise strategy games environment is totally new and the number of possible moves is almost infinite. Any brute-force-like methods will fail or require a lot of computational time to obtain any viable results, while common human player requires such an AI technique to perform actions almost instantly.

The research is now conducted on methods that would improve static, scripted AI, without significant effect on the speed of decision making. One of the popular ways seems to be "Reinforcement learning" [10]. This method can be used to optimize single decision made during the game using historical data. In [11] usage of this method to optimize algorithm for city placement in popular game *Civilization IV* is described, while in [3] optimization of high-level strategies for that game is presented. Those methods help improve AI efficiency a bit, but it is obvious that this is only tip of the iceberg. Both of the above mentioned research works focus on optimization of only one aspect of the game. Although the second approach allows for focusing on global strategy that impacts the entire game, it is still just choosing from four scripted tactics.

Another method is called "Dynamic Scripting" [9]. It is quite similar to reinforcement learning—it is focused on adjusting factors of scripts on the basis of historical data collected from previous runs of those scripts and thus improving their efficiency.

Such methods allow for AI improvement with the use of data coming from competition against other AI players, however the most important data is collected when playing against human opponents. In order to improve the learning process during competition with human players "online learning" can be used. It allows to collect the data generated directly by human players, accumulate it and adjust strategy of individual AI players.

While those methods are having their successes, the main problem is balance between exploration and exploitation. Exploitation of collected knowledge is needed to perform the best actions, while exploration is necessary to collect more data. So such approach always requires significant number of tries, before the results can be visible—and this is the situation when only one aspect of the game is taken into consideration. The more aspects are automated by these algorithms, the more tries are needed to produce a reasonable version of the AI technique for the given game. This is the main factor why the game producers do not like such approach. Usually everybody wants a product which is immediately ready for usage—not after some time (which is even hard to predict) needed for learning.

The key difference with the algorithm proposed in this paper is that in our technique adjustment of the strategy is done not on the basis of historical data, but it is based on simulations performed during the game, while trying to keep AI responses as quick as possible. Such an approach can be used in a strategy game to adjust several important decisions in the case of complex games, or even all of them in the case of less complex ones.

Algorithms like Monte-Carlo Tree Search (MCTS) are also used in games, however usually either to simulate simple actions, or in games that do not require quick response times, like Chess or Go. There are three main differences between the approach proposed in this paper, which is based on game tree search, and standard MCTS algorithm:

- Simulations of the moves are performed with the use of scripted AI algorithm, instead of using random movements.

- Simulations are not performed till the end of the game, but usually end after few rounds. This requires special formulas to calculate value of moves.
- Exploration versus exploitation balance is shifted in order to minimize the number of simulations needed to achieve reasonable outcome.

The above differences have one main purpose—minimize the time needed for performing simulations.

3 Puerto Rico Game

Puerto Rico is a German-style strategy board game, created by Andreas Seyfarth in 2002 and published by Alea [2]. The game quickly reached top ranks in BoardGameGeek (world's most popular board game site) and still stays in top10, becoming the oldest game at the top ranks [1].

There are three main reasons why we decided to focus our research on this game, and not on any other: **lack of randomness**, **level of rivalry** and **easy rules**. From all the board games we have played, Puerto Rico is less dependent on randomness, like throwing a dice or random events. It requires no luck, but pure tactical skills to win a game. That makes it a good test-bed for AI techniques. The level of rivalry in this game makes it impossible to create some unique "always good" strategy. Players fight for control over common pool of various kind of resources and performing good moves highly depends on opponent's moves. That additionally increases the difficulty level of creating AI technique properly reacting to environment changes (which always is a hard task). Easy rules decrease time spent on creating game engine and coding all possible situations on the board, allowing to focus on creating AI mechanisms.

In Puerto Rico game we can perform the following actions:

- Raise plantations, which later allows us to produce goods of 5 kinds: corn, indigo, sugar, tobacco and coffee.
- Build production buildings, which combined with plantations, let us produce goods.
- Build utility buildings, which gives us several bonuses.
- Grant colonists to work on plantations or buildings.
- Produced goods can either be sold on the Market, granting doubloons, or shipped to Europe, granting victory points.
- Every turn we pick one of the 6–7 roles. Then every player performs action connected with this role, however the player who picked it is granted additional bonus and also opportunity to perform the given role's action first. Possible roles are the following: Settler (raising plantations), Mayor (getting new colonists and transfer them to plantations/buildings), Builder (builds buildings for the doubloons), Craftsman (produces goods), Trader (sells goods for the doubloons), Captain (ships goods for victory points) and Prospector (no action).

4 Game Engine

The game engine used in this research was created in C# using MS Visual Studio, with .Net 4.5 and WPF frameworks. WPF allows to create a nice graphical interface for the game, while .Net features, especially generic, allows to create a clean object oriented code. The engine of the game consists of 3 main modules (classes): Game, GamePage and Player. First two ones are responsible for general game rules, while Player class holds all user actions. By default these are human player actions performed with the use of mouse. Every AI player class derives from Player class, overriding its virtual functions, which are responsible for making choices (there are 10 such functions). Such architecture makes it quite easy to add a new AI technique.

5 AI Techniques

Low complexity level of the game rules leads to only 10 different possible choices to be made. Of course these choices always appear within the context of different situation—giving almost infinite number of possible combinations—but it allows for easier grouping of all AI actions. The set of choices include:

- Choosing a role;
- Choosing a plantation from available plantations stack;
- Choosing a building to build from the list of available buildings;
- Choosing how to re-allocate colonists over plantations and buildings;
- Choosing an additional production (as a bonus to craftsman role);
- Choosing which good to sell on the Market;
- Choosing which type of goods should be loaded onto ship;
- Choosing a ship onto which the goods should be loaded;
- Choosing which goods will be spoiled (due to the lack of storage);
- Choosing whether to use a Hacienda (an utility building).

The above grouping allowed for creating 10 virtual functions, one for each choice. Initially these functions were created for a human player and later they were overridden in AI classes, which provided specific implementation.

For a human player, a specific GUI is launched, which allows for making a choice with a mouse click. For AI players it is required to create a specific algorithm for each of those functions.

Two kinds (generations) of AI mechanisms were created for the Puerto Rico game. First one is similar to the already mentioned scripted AI (with some small adjustments implemented). The second one does not have any hard-coded strategies—instead it tries to perform a simulation for choosing best moves.

5.1 Scripted AI

First generation of AI was based mostly on scripts and Bayesian networks, with a little addition of random algorithms.

Most of the 10 possible actions, mentioned above, have a very similar algorithm:

1. Prepare a list of all possible choices (e.g. all buildings that are possible to build taking into account the amount of money and placement limitations).
2. Calculate which choice is potentially the best one by adding specific weights to each of them and then select one of the possibilities on the basis of these weights.
   - Add an a priori weight, which is based on some values predefined for each of the choices.
   - Consider multiple cases that can modify weights accordingly to the current situation on the board.
3. Randomly choose final option, on the basis of weighted list (choice with highest weight is the most probable).

Slight changes had to be made for re-allocating colonists action. Due to the fact the we have to place multiple colonists, the above algorithm actually runs in a loop and the choice is done multiple times, until all colonists are placed.

It should be noticed that there is a slight difference when we compare our approach to the standard scripted approach—it is randomness factor. It was added to achieve the lack of predictability, so AI player not always acts in the same way in a similar situation (e.g. at the beginning of the game). Puerto Rico game offers multiple, almost equally valuable, choices at many stages of the game, so randomness factor is used in order to avoid following the same path in the case when there are several reasonable possibilities. In order to avoid the selection of bad moves, precalculated weights were added, so it is very unlikely that really bad (low value) moves would be chosen.

Such AI approach is extremely fast, relatively easy to code and slight randomness factor gives lower rates of predictability. The above approach has also some limitations. It does not "think"—all the decisions are defined by its creator (so basically it acts like him in some limited way). Quick (not very sophisticated) implementation of such approach would result in quite low strategy skills, which would not be really challenging for advanced human player. This approach needs a lot of work when trying to implement large number of cases, resulting in performing better moves in different situations and generally being more challenging for advanced players. Randomness factor, which protects against the predictability, unfortunately sometimes lowers general skills of AI player, so it should be used with caution.

5.2 Thinking AI

The second generation of AI is based on game tree searching approach. During the tree search, simulations of moves are performed with the use of first generation (scripted) AI. In short, the algorithm works as follows:

1. When the choice is to be made, start a simulation. To save some time, start it in the background as soon as possible (e.g. we can start thinking about which building should be built, as soon as other player picks the Builder role).
2. Like previously, create a list of all possible choices.

3. Start a simulation:
   - Create a copy of the game with the current state.
   - Replace all the players with Scripted AI.
   - Make a choice (pick one of possible choices from the list).
   - Let the game run for a turn or few turns (depending on the type of choice).
   - Estimate the final result. Estimation is different depending on the choice. For example if it is about choosing a resource to be sold on the Market then the estimation is quite simple—the amount of money earned is compared to the amount of money earned by other players. If it is about a role to choose then player's virtual score after few rounds of the game is estimated. This estimation is made on the basis of goods, money and holdings, as well as on current victory points.
4. Repeat a simulation multiple times for each choice and prepare the statistics.
5. When it is time to make a choice, stop the simulation and choose statistically best action.

Again, there is a slight difference for re-allocating colonists action. To save some time, most of the colonists are placed with the use of scripted algorithm (because usually it is an obvious choice), only for few last colonists simulation is made.

Choosing a Role may be implemented as follows. First, for every possible choice 5 rounds of the game are played. Next the score = $\sum_{P_i} \text{Difference}(P_i)$ is calculated, where *P(i)* are all other players in the game.

Difference is calculated as follows:

$$\text{Difference} = (Pr + 20) \cdot (VP0 - VP_i) + (80 - Pr) \cdot (Pos0 - Pos_i) \quad (1)$$

where Pr is a game progress valued from 0 to 100, $VP0$ are Victory Points of AI, VP_i are Victory Points of i-th player, $Pos0$ is Virtual Score (based on possession state) of AI and Pos_i is Virtual Score (based on possession state) of i-th player.

Possession state is calculated in the following way:

$$\begin{aligned}\text{Pos} = (100 \cdot D + 50 \cdot C + \sum_{R_i} \text{Quan}(R_i)(100 + 50 \cdot \text{Val}(R_i) + \text{Prod}(R_i) \cdot (200+ \\ 100 \cdot \text{Val}(R_i)) + 450 \cdot Q + \sum_{Pl_i} 50 + 25 \cdot \text{Val}(Pl_i) + \sum_{B_i} 150 \cdot \text{Prize}(B_i)\end{aligned} \quad (2)$$

where D is Quantity of doubloons, C is Quantity of colonists, R_i is type of resource, Quan is Quantity of resource, Val is Value of type of the resource (Corn = 1, Indigo = 2, Sugar = 3, Tobacco = 4, Coffee = 5), Prod is Production capacity of the resource, Q is Quantity of Quarries, Pl_i is i-th plantation in possession, B_i is i-th building in possession and Prize is prize for building.

It is worth to notice (Eq. (1)) that game progress is very important. In early stages of the game it is more important to have high Virtual Score (resources that will allow to produce Victory Points later), while later it is more important to have Victory Points.

There are of course some pros and cons of such AI technique. Firstly, we can say that this AI is "thinking", choices are not coded by the creator—instead they are the results of simulations. Implementation of this algorithm gives good results and do not require a lot of work. This approach possesses a highly challenging strategy skills, capable of competing with experienced human player. It can dynamically react to the changes taking place on the board and adjust accordingly its tactics. Its actions and tactics are very hard to predict.

The biggest disadvantage of the second generation AI is that it is dependent on other (scripted) AI, it simply can not work without it. And actually its skills also depend on how precisely the scripted AI was developed. But even with quite basic implementation of the first generation AI, the second generation is working very well. The time needed for performing moves is noticeable, but not annoying for human players. This technique, when given more time for performing a move, gives slightly better results, but on any modern PC quite short time is enough for very good performance (it is usually between one and several seconds for a single action). Implementing estimation functions—which are most important for making a proper choice—can be quite difficult. The functions can be obtained by a series of additional simulations or by applying reinforcement learning technique, which seems to be a next step to be made in order to further improve AI's efficiency.

6 Experiments and Results

6.1 Comparison of Both AI Mechanisms

Assumption of the research was that Thinking AI should be a perfect opponent, able to win with Scripted AI—at least if given enough time for thinking. "Thinking time" is a maximum time dedicated for performing simulations by Thinking AI. Some of the functions use the whole time, some only 1/2 or 1/4 of a given time. A series of experiments were carried out in order to test that. The configuration details of the machine used during experiments are as follows: Intel i5-2500 k CPU, 3.30 GHz, 16 GB RAM running MS Windows 7. Rules of the experiments were as follows:

- Run tests for 4 possible settings of thinking time: 100 ms, 2.5 s, 5 s,10 s.
- Run 100 games for each series, with 4 players, 2 of each AI types.
- Check the game winner and prepare statistics based on that.

Results (see Fig. 1) were a bit surprising. It could be expected that with minimal thinking time Scripted AI would win because Thinking AI would perform quite random moves, but in reality it was opposite. Apparently 100 ms was enough to perform such a number of simulations that Scripted AI could be outplayed. With 2.5 s thinking time the second generation AI almost always won.

6.2 AI Vs Human Players

Experiments in which AI competed with human players were much more difficult to carry out. During those experiments, one of the authors of this paper and few other players with various range of strategy skills played about 50 games.

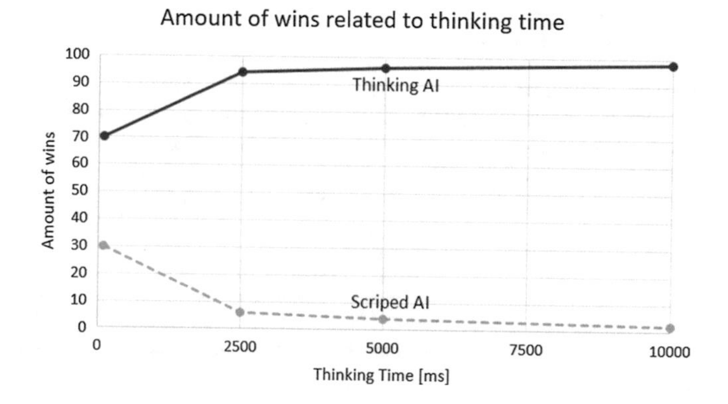

Fig. 1. Comparison of AI mechanisms

Table 1. Results of competition between AIs and expert player

Player	Score in consecutive games																				Average score
AI Scripted 1	1	2	1	1	2	1	2	3	1	2	2	2	1	3	1	1	2	1	1	1	1.55
AI Scripted 2	2	1	2	2	1	2	1	1	2	1	3	1	2	1	2	2	1	2	2	2	1.65
AI Thinking 1	4	5	4	3	4	5	4	2	3	5	5	5	3	4	3	4	3	5	4	5	4
AI Thinking 2	3	3	5	5	3	4	3	4	5	4	4	3	4	2	5	5	5	4	3	4	3.9
Expert player	5	4	3	4	5	3	5	5	4	3	1	4	5	5	4	3	4	3	5	3	3.9

Most of the games were played by an expert player against 2 AIs of both types. In these games each player received points on the basis of his position at the end of the game (5 points for winning, 4 for second place, etc.) Then statistical data was computed on the basis of the results of 20 games in order to compare performance of human expert player during competition against different AIs (see Table 1). As we can see, Scripted AI had no chance at all, while Thinking AI scored almost the same as expert player. Statistics from the games played against beginner or intermediate human player are currently not available because the focus of this research was to create the best possible AI player capable of competing with an expert human player. However a few games were played also against less advanced human players and typical results of one of such games are presented in the Fig. 2.

The following observations may be formulated on the basis of the obtained results. Scripted AI is a good opponent for beginning player or a player with lower strategy skills. Intermediate player will often win with the Scripted AI, but winning against Thinking AI will be almost impossible. Human player needs to possess good strategy skills to be able to win with it at all. Advanced player is usually capable of beating the first generation AI and sometimes the second generation AI. Only very good player would win with Thinking AI more often, but still not always. On the basis of the performed experiments it is safe to assume that Thinking AI's skills are similar to such an experienced human player.

Game ended after 19 rounds!

Ada (AI II) - 67 victory points (export: 32, building: 20, special: 15), 6 doubloons, 12 goods
Ela (AI II) - 63 victory points (export: 30, building: 20, special: 13), 3 doubloons, 6 goods
Player 1 - 47 victory points (export: 33, building: 14, special: 0), 3 doubloons, 13 goods
Zosia (AI I) - 35 victory points (export: 26, building: 9, special: 0), 4 doubloons, 5 goods
Maniek (AI I) - 34 victory points (export: 13, building: 21, special: 0), 11 doubloons, 4 goods

The winner is Ada (AI II). Congratulations!

Fig. 2. Typical results of competition between intermediate human player and various AIs (*AI I* is Scripted AI, *AI II* is Thinking AI)

7 Summary and Conclusions

The AI players proposed in this paper were, more or less, performing in accordance with our expectations. Two types of AI, with different playing styles and skills, can ensure a good game experience both for beginner and advanced human players. It is really hard to get bored when playing against such AI players—actions of both of them are rather hard to predict. Especially Thinking AI is very promising technique because it can adapt and dynamically react to other players' actions.

It is worth to notice that there are still possibilities to improve both AI mechanisms. Improving Scripted AI relies on tests and observation. We can try to improve conditional statements and weights calculations used during decision making. This technique can be also greatly improved by adding more conditional statements, taking into account additional situations. Possibly, also dynamic scripting could be used here, however it would require much more programming effort.

Improving Thinking AI can be done automatically when improving Scripted AI. Improvement of Scripted AI would impact simulation results because moves would be performed against stronger opponents. Another way to improve Thinking AI is to create better formulas for approximation of the best moves. The most important thing would be more precise approximation of a real value of opponent player's assets. It could be done by performing additional simulations or by applying reinforcement learning technique. The goal would be to find better ways of valuing different kind of opponent players' properties, for example how to value buildings, plantations and colonists as compared to doubloons and victory points. Creating better formula could eventually lead to choosing much better moves from the simulated ones.

The proposed Thinking AI technique, of course when correctly implemented, seems to be highly effective and significantly improving skills of AI player. Due to the fact that it is based on the standard scripted AI, it could be used along with other known techniques (that improve efficiency of the scripted AI) and thus made even more challenging for human players.

In the future research we plan also to adapt and use the techniques proposed in this paper in agent-based strategy and tactical games, where agents are individuals interacting independently with each other and with an environment. In such a case AI mechanisms are encapsulated within each agent and thus many different AI techniques can be coherently combined together (for example evolutionary algorithms, neural nets, multi-objective evolutionary optimization techniques, etc.)—we have already used such approach with great success in computational systems [4–7] and it seems that it can also bring many advantages to the construction of AI for computer games. The simulation technique proposed in this paper could be used by each agent in order to assess its future moves and to select the best one.

Acknowledgments. This research was partially supported by Polish Ministry of Science and Higher Education under AGH University of Science and Technology, Faculty of Computer Science, Electronics and Telecommunications statutory project no. 11.11.230.124.

References

1. BoardGameGeek. https://boardgamegeek.com/browse/boardgame
2. Puerto Rico (board game). https://en.wikipedia.org/wiki/Puerto_Rico_(board_game)
3. Amato, C., Shani, G.: High-level reinforcement learning in strategy games. In: van der Hoek, W. et al. (ed.) Proceedings of 9th International Conference on Autonomous Agents and Multiagent Systems (AAMAS 2010), pp. 75–82. IFAAMAS (2010)
4. Cetnarowicz, K., Dreżewski, R.: Maintaining functional integrity in multi-agent systems for resource allocation. Comput. Inform. **29**(6), 947–973 (2010)
5. Dreżewski, R., Sepielak, J.: Evolutionary system for generating investment strategies. In: Giacobini, M., et al. (eds.) Applications of Evolutionary Computing, EvoWorkshops 2008. LNCS, vol. 4974, pp. 83–92. Springer, Heidelberg (2008)
6. Dreżewski, R., Siwik, L.: Multi-objective optimization technique based on co-evolutionary interactions in multi-agent system. In: Giacobini, M., et al. (eds.) Applications of Evolutionary Computing, EvoWorkshops 2007. LNCS, vol. 4448, pp. 179–188. Springer, Heidelberg (2007)
7. Dreżewski, R., Siwik, L.: Co-evolutionary multi-agent system for portfolio optimization. In: Brabazon, A., O'Neill, M. (eds.) Natural Computing in Computational Finance, vol. 1, pp. 271–299. Springer, Heidelberg (2008)
8. Millington, I., Funge, J.: Artificial Intelligence for Games. CRC Press, Boston (2009)
9. Spronck, P., Ponsen, M.J.V., Sprinkhuizen-Kuyper, I.G., Postma, E.O.: Adaptive game AI with dynamic scripting. Mach. Learn. **63**(3), 217–248 (2006)
10. Szepesvári, C.: Algorithms for Reinforcement Learning. Synthesis Lectures on Artificial Intelligence and Machine Learning. Morgan & Claypool Publishers (2010)
11. Wender, S., Watson, I.: Using reinforcement learning for city site selection in the turn-based strategy game Civilization IV. In: Hingston, P., Barone, L. (eds.) CIG, pp. 372–377. IEEE (2008)

Simple Bounded MTLK Model Checking for Timed Interpreted Systems

Agnieszka M. Zbrzezny(✉) and Andrzej Zbrzezny

IMCS, Jan Długosz University,
Al. Armii Krajowej 13/15, 42-200 Częstochowa, Poland
{agnieszka.zbrzezny,a.zbrzezny}@ajd.czest.pl

Abstract. We present a new translation of Metric Temporal Logic with knowledge operators (MTLK) to the Linear Temporal Logic with knowledge operators and with a new set of the atomic propositions ($\mathrm{LTL_qK}$). We investigate a SAT-based bounded model checking (BMC) method for MTLK. The semantics of MTLK is defined over timed interpreted systems (TIS). We show how to implement the bounded model checking technique for $\mathrm{LTL_qK}$ and timed interpreted systems, and as a case study, we apply the technique in the analysis of the Timed Generic Pipeline Paradigm modelled by TIS. We also present the differences between the old translation of MTLK and the new one. The theoretical description is supported by the experimental results that demonstrate the efficiency of the method.

1 Introduction

The formalism of *interpreted systems* (IS) was introduced in [2] to model multi-agent systems (MAS) [5], which are intended for reasoning about the agents' epistemic and temporal properties. The formalism of timed interpreted systems (TIS) [6] extends IS to make the reasoning possible about not only temporal and epistemic properties, but also about real-time aspects of MASs.

Bounded model checking [1,4] (BMC) is one of the symbolic model checking technique designed for finding witnesses for existential properties or counterexamples for universal properties. Its main idea is to consider a model reduced to a specific depth, which means that we consider only finite prefixes of the paths in the model. The SAT-BMC method works by mapping a bounded model checking problem to the satisfiability problem (SAT). For metric temporal logic with epistemic operators (MTL) [3,6] and timed interpreted systems [6] the BMC method can by described as follows: given a model $\mathcal{M}$ for a timed interpreted system, an MTLK formula φ, and a bound k, a model checker creates a propositional formula $[\mathcal{M}, \varphi]_k$ that is satisfiable if and only if the formula φ is true in the model $\mathcal{M}$. The novelty of our paper lies in:

Partly supported by National Science Centre under the grant No. 2014/15/N/ST6/05079.

G. Jezic et al. (eds.), *Agent and Multi-Agent Systems: Technology and Applications*, Smart Innovation, Systems and Technologies 74,
DOI 10.1007/978-3-319-59394-4_9

1. defining a translation of the existential model checking problem for MTLK to the existential model checking problem for linear temporal logic with additional propositional variables q_I. This logic is denoted by LTL_qK;
2. defining bounded semantics for LTL_qK and defining the BMC algorithm;
3. implementing the new method.

The translation from MTLK to LTL_qK requires neither new clocks nor new transitions, whereas the translation to HLTLK [6] requires as many new clocks as there are intervals in a given formula. It also requires an exponential number of resetting transitions. Moreover, our BMC method needs only one path for temporal operators, whereas the BMC method from [6] needs a number of paths depending on a given formula φ. Thus, one may expect that our method is much more effective since intuition is that an encoding which results in fewer variables and clauses is usually easier to solve.

Finally, we evaluate the BMC method experimentally by means of a timed generic pipeline paradigm (TGPP), which we model by a TIS.

The rest of the paper is structured as follows. In Sect. 2 we briefly recall the basic notion used through the paper. In Sect. 3 we define the translation to LTL_qK. In Sect. 4 we define the BMC method for LTL_qK. In Sect. 5 we discuss our experimental results. In Sect. 6 we conclude the paper.

2 Preliminaries

Let $\mathbb{N}$ be a set of natural numbers. We assume a finite set $\mathcal{X} = \{x_0, \ldots, x_{n-1}\}$ of variables, called *clocks*. Each clock is a variable ranging over a set of non-negative natural numbers. A *clock valuation* is a total function $v : \mathcal{X} \mapsto \mathbb{N}$ that assigns to each clock x a non-negative integer value $v(x)$. The set of all the clock valuations is denoted by $\mathbb{N}^n$. For $X \subseteq \mathcal{X}$, the valuation $v' = v[X := 0]$ is defined as: $\forall x \in X$, $v'(x) = 0$ and $\forall x \in \mathcal{X} \setminus X$, $v'(x) = v(x)$. For $\delta \in \mathbb{N}$, $v + \delta$ denotes the valuation v'' such that $\forall x \in \mathcal{X}, v''(x) = v(x) + \delta$. Let $x \in \mathcal{X}$, $c \in \mathbb{N}$, and $\sim\, \in \{<, \leqslant, =, \geqslant, >\}$. The set $\mathcal{C}(\mathcal{X})$ of *clock constraints* over the set of clocks $\mathcal{X}$ is defined by the following grammar: $\mathfrak{cc} := x \sim c \mid \mathfrak{cc} \wedge \mathfrak{cc}$. Let v be a clock valuation, and $\mathfrak{cc} \in \mathcal{C}(\mathcal{X})$. A clock valuation v satisfies a clock constraint $\mathfrak{cc}$, written as $v \models \mathfrak{cc}$, iff $\mathfrak{cc}$ evaluates to true using the clock values given by the valuation v.

Let $\mathcal{A} = \{1, \ldots, n\}$ denote a non-empty and finite set of agents, and $\mathcal{E}$ be a special agent that is used to model the environment in which the agents operate, and $\mathcal{AP} = \bigcup_{\mathbf{i} \in \mathcal{A} \cup \{\mathcal{E}\}} \mathcal{AP}_{\mathbf{i}}$ be a set of atomic formulae, such that $\mathcal{AP}_{\mathbf{i}_1} \bigcap \mathcal{AP}_{\mathbf{i}_2} = \emptyset$ for all $\mathbf{i}_1, \mathbf{i}_2 \in \mathcal{A} \cup \{\mathcal{E}\}$.

A *timed interpreted system* is a tuple $(\{L_{\mathbf{i}}, Act_{\mathbf{i}}, \mathcal{X}_{\mathbf{i}}, P_{\mathbf{i}}, \mathcal{V}_{\mathbf{i}}, Inv_{\mathbf{i}}, \iota_{\mathbf{i}}\}_{\mathbf{i} \in \mathcal{A} \cup \{\mathcal{E}\}}, \{t_{\mathbf{i}}\}_{\mathbf{i} \in \mathcal{A}}, \{t_{\mathcal{E}}\})$, where: $L_{\mathbf{i}}$ is a non-empty set of *locations* of the agent $\mathbf{i}$, $\iota_{\mathbf{i}} \subseteq L_{\mathbf{i}}$ is a non-empty set of initial locations, $Act_{\mathbf{i}}$ is a non-empty set of *possible actions* of the agent $\mathbf{i}$, $Act = Act_1 \times \ldots \times Act_n \times Act_{\mathcal{E}}$ is the set of *joint actions*, $\mathcal{X}_{\mathbf{i}}$ is a non-empty set of *clocks*, $P_{\mathbf{i}} : L_{\mathbf{i}} \rightarrow 2^{Act_{\mathbf{i}}}$ is a *protocol function* modelling the program the agent is executing. Formally, for any agent $\mathbf{i}$, the actions of the agents are selected according to a local protocol, $t_{\mathbf{i}} : L_{\mathbf{i}} \times L_{\mathcal{E}} \times \mathcal{C}(\mathcal{X}_{\mathbf{i}}) \times 2^{\mathcal{X}_{\mathbf{i}}} \times Act \rightarrow L_{\mathbf{i}}$ is a (partial) *evolution function* for agents. The evolution function determines how locations "evolve", based on the agent's locations, on other agents' actions, on the location of a special agent used to model the environment, on the clock constraints of agent $\mathbf{i}$, and on the set of clocks, $t_{\mathcal{E}} : L_{\mathcal{E}} \times \mathcal{C}(\mathcal{X}_{\mathcal{E}}) \times 2^{\mathcal{X}_{\mathcal{E}}} \times Act \rightarrow L_{\mathcal{E}}$ is a (partial) *evolution function* for environment, $\mathcal{V}_{\mathbf{i}} : L_{\mathbf{i}} \rightarrow 2^{\mathcal{AP}_{\mathbf{i}}}$ is a *valuation function* assigning to each location a set of atomic formulae that are assumed to be true at that location, $Inv_{\mathbf{i}}$: $L_{\mathbf{i}} \rightarrow \mathcal{C}(\mathcal{X}_{\mathbf{i}})$ is an *invariant function*, that constraints the amount of time the agent $\mathbf{i}$ may spend in a given location.

It is assumed that locations, actions and clocks for the environment are "public", which means that all the agents know the current location, the action, and the clock valuation of the environment.

We also assume that if $\epsilon_{\mathbf{i}} \in P_{\mathbf{i}}(\ell_{\mathbf{i}})$, then $t_{\mathbf{i}}(\ell_{\mathbf{i}}, \ell_{\mathcal{E}}, \mathfrak{cc}_{\mathbf{i}}, \mathcal{X}, (a_1, \ldots, a_n, a_{\mathcal{E}})) = \ell_{\mathbf{i}}$ for $a_{\mathbf{i}} = \epsilon_{\mathbf{i}}$, any $\mathfrak{cc}_{\mathbf{i}} \in \mathcal{C}(\mathcal{X}_{\mathbf{i}})$, and any $\mathcal{X} \subseteq \mathcal{X}_{\mathbf{i}}$. Each element t of $t_{\mathbf{i}}$ is denoted by $< \ell_{\mathbf{i}}, \ell_{\mathcal{E}}, \mathfrak{cc}_{\mathbf{i}}, \mathcal{X}', a, \ell'_{\mathbf{i}} >$, where $\ell_{\mathbf{i}}$ is the source location, $\ell'_{\mathbf{i}}$ is the target location, a is an action, $\mathfrak{cc}_{\mathbf{i}}$ is the enabling condition for $t_{\mathbf{i}}$, and $\mathcal{X}' \subseteq \mathcal{X}_{\mathbf{i}}$ is the set of clocks to be reset after performing t. An invariant condition allows the TIS to stay at the location ℓ as long as the constraint $Inv_{\mathbf{i}}(\ell_{\mathbf{i}})$ is satisfied. The guard $\mathfrak{cc}$ has to be satisfied to enable the transition.

Timed model. For a given TIS let the symbol $S = \prod_{\mathbf{i} \in \mathcal{A} \cup \{\mathcal{E}\}} (L_{\mathbf{i}} \times \mathbb{N}^{\mathcal{X}_{\mathbf{i}}})$ denote the non-empty set of all the *global states*. Moreover, for a given global state $s = ((\ell_1, v_1), \ldots, (\ell_n, v_n), (\ell_{\mathcal{E}}, v_{\mathcal{E}})) \in S$, let the symbols $l_{\mathbf{i}}(s) = \ell_{\mathbf{i}}$ and $v_{\mathbf{i}}(s) = v_{\mathbf{i}}$ denote, respectively, the local component and the clock valuation of agent $\mathbf{i} \in \mathcal{A} \cup \{\mathcal{E}\}$ in s. Now, for a given TIS we define a *timed model* (or a *model*) as a tuple $\mathcal{M} = (S, Act, \iota, T, \mathcal{V})$, where:

$Act = Act_1 \times \ldots \times Act_n \times Act_{\mathcal{E}}$ is the set of all the joint actions, $\iota = \prod_{\mathbf{i} \in \mathcal{A} \cup \{\mathcal{E}\}} (\iota_i \times \{0\}^{\mathcal{X}_i})$ is the set of all the *initial* global states, $\mathcal{V} : S \rightarrow 2^{\mathcal{AP}}$ is the valuation function defined as $\mathcal{V}(s) = \bigcup_{\mathbf{i} \in \mathcal{A} \cup \{\mathcal{E}\}} \mathcal{V}_{\mathbf{i}}(l_{\mathbf{i}}(s))$, $T \subseteq S \times (Act \cup \mathbb{N}) \times S$ is a transition relation defined by action and time transitions. Let s and s' be two global states. For $\widetilde{a} \in Act$:

1. action transition: $(s, \widetilde{a}, s') \in T$ (or $s \xrightarrow{\widetilde{a}} s'$) iff for all $\mathbf{i} \in \mathcal{A} \cup \{\mathcal{E}\}$, there exists a local transition $t_{\mathbf{i}}(l_{\mathbf{i}}(s), \mathfrak{cc}_{\mathbf{i}}, \mathcal{X}', \widetilde{a}) = l_{\mathbf{i}}(s')$ such that $v_{\mathbf{i}}(s) \models \mathfrak{cc}_{\mathbf{i}} \wedge Inv(l_{\mathbf{i}}(s))$ and $v'_{\mathbf{i}}(s') = v_{\mathbf{i}}(s)[\mathcal{X}' := 0]$ and $v'_{\mathbf{i}}(s') \models Inv(l_{\mathbf{i}}(s'))$ ($v_{\mathbf{i}}(s)[\mathcal{X}' := 0]$ denotes the clock valuation which assigns 0 to each clock in $\mathcal{X}'$ and agrees with $v_{\mathbf{i}}(s)$ over the rest of the clocks).
2. time transition: let $\delta \in \mathbb{N}$, $(s, \delta, s') \in T$ iff for all $\mathbf{i} \in \mathcal{A} \cup \mathcal{E}$, $l_{\mathbf{i}}(s) = l_{\mathbf{i}}(s')$ and $v_{\mathbf{i}}(s) \models Inv(l_{\mathbf{i}}(s))$ and $v'_{\mathbf{i}}(s') = v_{\mathbf{i}}(s) + \delta$ and $v'_{\mathbf{i}}(s') \models Inv(l_{\mathbf{i}}(s'))$.

Given a TIS, one can define for any agent $\mathbf{i}$ the indistinguishability relation $\sim_{\mathbf{i}} \subseteq S \times S$ as follows: $s \sim_{\mathbf{i}} s'$ iff $l_{\mathbf{i}}(s') = l_{\mathbf{i}}(s)$ and $v_{\mathbf{i}}(s') = v_{\mathbf{i}}(s)$. We assume the following definitions of epistemic relations: $\sim^E_\Gamma \stackrel{def}{=} \bigcup_{\mathbf{i} \in \Gamma} \sim_{\mathbf{i}}$, $\sim^C_\Gamma \stackrel{def}{=} (\sim^E_\Gamma)^+$ (the transitive closure of $\sim^E_\Gamma$), $\sim^D_\Gamma \stackrel{def}{=} \bigcap_{\mathbf{i} \in \Gamma} \sim_{\mathbf{i}}$, where $\Gamma \subseteq \mathcal{A}$.

A run in $\mathcal{M}$ is an infinite sequence $\rho = s_0 \stackrel{\delta_0,a_0}{\longrightarrow} s_1 \stackrel{\delta_1,a_1}{\longrightarrow} s_2 \stackrel{\delta_2,a_2}{\longrightarrow} \ldots$ of global states such that the following conditions hold for all $i \in \mathbb{N} : s_i \in S, a_i \in Act, \delta_i \in \mathbb{N}_+$, and there exists $s_i' \in S$ such that $(s_i, \delta_i, s_i') \in T$ and $(s_i, a_i, s_{i+1}) \in T$.

Observe that the above definition of the run ensures that the first transition is the time one, and between each two action transitions at least one time transition appears.

The set of all the runs starting at $s \in S$ is denoted by $\Pi(s)$, and the set of all the runs starting at an initial state is denoted by $\Pi = \bigcup_{s^0 \in \iota} \Pi(s^0)$. Moreover, for $\widetilde{a} \in Act \cup \{\tau\}$, we sometimes write $s \stackrel{\widetilde{a}}{\longrightarrow} s'$ instead of $(s, \widetilde{a}, s') \in T$.

MTLK. Let $p \in \mathcal{AP}$, and I be an interval in $\mathbb{N}$ of the form: $[a, b)$ or $[a, \infty)$, for $a, b \in \mathbb{N}$ and $a \neq b$. The MTLK in negation normal form is defined by the following grammar:

$$\alpha := \textbf{true} \mid \textbf{false} \mid p \mid \neg p \mid \alpha \wedge \alpha \mid \alpha \vee \alpha \mid \alpha \mathbf{U}_{\mathrm{I}} \alpha \mid \mathbf{G}_{\mathrm{I}} \alpha \mid \mathrm{K}_{\mathbf{i}} \varphi \mid \overline{\mathrm{K}}_{\mathbf{i}} \varphi.$$

Intuitively, $\mathbf{U}_{\mathrm{I}}$ and $\mathbf{G}_{\mathrm{I}}$ are the operators for *bounded until* and for *bounded always*. The formula $\alpha \mathbf{U}_{\mathrm{I}} \beta$ is true in a computation if β is true in the interval I at least in one state and always earlier α holds. The formula $\mathbf{G}_{\mathrm{I}} \alpha$ is true in a computation if α is true at all states of the computation that are in the interval I. The derived basic modality is defined as follows: $\mathbf{F}_{\mathrm{I}} \alpha \stackrel{def}{=} \textbf{true} \mathbf{U}_{\mathrm{I}} \alpha$ (*bounded eventually*). $\overline{\mathrm{K}}_{\mathbf{i}}$ is the operator dual for the standard epistemic modality $\mathrm{K}_{\mathbf{i}}$ ("agent **i** knows"), so $\overline{\mathrm{K}}_{\mathbf{i}} \alpha$ is read as "agent **i** does not know whether or not α holds".

The EMTLK in the existential fragment of MTLK defined as:

$$\alpha := \textbf{true} \mid \textbf{false} \mid p \mid \neg p \mid \alpha \wedge \alpha \mid \alpha \vee \alpha \mid \alpha \mathbf{U}_{\mathrm{I}} \alpha \mid \mathbf{G}_{\mathrm{I}} \alpha \mid \overline{\mathrm{K}}_{\mathbf{i}} \varphi.$$

Observe that we assume that MTLK (and so EMTLK) formulae are given in the negation normal form, in which the negation can be only applied to propositional variables. Moreover, EMTLK is existential only w.r.t. the epistemic modalities.

In order to define the satisfiability relation for MTLK, we need to define the notion of a *discrete path λ_ρ corresponding to run ρ* [6]. This can be done in a unique way because of the assumption that $\delta_i \in \mathbb{N}_+$. First, define the sequence $\Delta_0 = [b_0, b_1), \Delta_1 = [b_1, b_2), \Delta_2 = [b_2, b_3), \ldots$ of pairwise disjoint intervals, where: $b_0 = 0$, and $b_i = b_{i-1} + \delta_{i-1}$ if $i > 0$. Now, for each $t \in \mathbb{N}$, let $idx_\rho(t)$ denote the unique index i such that $t \in \Delta_i$. A *discrete path* (or *path*) λ_ρ *corresponding to* ρ is a mapping $\lambda_\rho : \mathbb{N} \mapsto S$ such that $\lambda_\rho(t) = (\ell_i, v_i + t - b_i)$, where $i = idx_\rho(t)$. Given $t \in \mathbb{N}$, the suffix λ_ρ^t of a path λ_ρ at time t is a path defined as: $\forall i \in \mathbb{N}$, $\lambda_\rho^t(i) = \lambda_\rho(t + i)$.

In order to improve readability, in the following definition we write $\lambda_\rho^t \models_{\mathrm{MTLK}} \varphi$ instead of $\widehat{\mathcal{M}}, \lambda_\rho^t \models_{\mathrm{MTLK}} \varphi$, for any MTL formula φ.

Definition 1. *The* satisfiability *relation* $\models_{MTLK}$, *which indicates truth of an* MTLK *formula in the model* $\mathcal{M}$ *along a path* λ_ρ *at time* $t \in \mathbb{N}$, *is defined inductively as follows:*

- $\lambda_\rho^t \models_{\mathrm{MTLK}}$ **true**, $\lambda_\rho^t \not\models_{\mathrm{MTLK}}$ **false**,
- $\lambda_\rho^t \models_{\mathrm{MTLK}} p$ iff $p \in \mathcal{V}(\lambda_\rho(t))$, $\lambda_\rho^t \models_{\mathrm{MTLK}} \neg p$ iff $p \notin \mathcal{V}(\lambda_\rho(t))$,
- $\lambda_\rho^t \models_{\mathrm{MTLK}} \alpha \wedge \beta$ iff $\lambda_\rho^t \models_{\mathrm{MTLK}} \alpha$ and $\lambda_\rho^t \models_{\mathrm{MTLK}} \beta$,
- $\lambda_\rho^t \models_{\mathrm{MTLK}} \alpha \vee \beta$ iff $\lambda_\rho^t \models_{\mathrm{MTLK}} \alpha$ or $\lambda_\rho^t \models_{\mathrm{MTLK}} \beta$,
- $\lambda_\rho^t \models_{\mathrm{MTLK}} \alpha \mathbf{U}_{\mathrm{I}} \beta$ iff $(\exists t' \in \mathrm{I})(\lambda_\rho^{t+t'} \models_{\mathrm{MTLK}} \beta$ and $(\forall 0 \leqslant t'' < t')\lambda_\rho^{t+t''} \models_{\mathrm{MTLK}} \alpha)$,
- $\lambda_\rho^t \models_{\mathrm{MTLK}} \mathbf{G}_{\mathrm{I}} \beta$ iff $(\forall t' \in \mathrm{I})(\lambda_\rho^{t+t'} \models_{\mathrm{MTLK}} \beta)$,
- $\lambda_\rho^t \models_{\mathrm{MTLK}} \mathrm{K}_{\mathbf{i}} \alpha$ iff $(\forall \pi' \in \Pi)(\forall i \geq 0)(\pi'(i) \sim_{\mathbf{i}} \pi(t)$ implies $\mathcal{M}, \pi'^i \models \alpha)$,
- $\lambda_\rho^t \models_{\mathrm{MTLK}} \overline{\mathrm{K}}_{\mathbf{i}} \alpha$ iff $(\exists \pi' \in \Pi)(\exists i \geq 0)(\pi'(i) \sim_{\mathbf{i}} \pi(t)$ and $\mathcal{M}, \pi'^i \models \alpha)$.

As $\lambda_\rho^0 = \lambda_\rho$, we shall write $\mathcal{M}, \lambda_\rho \models_{\mathrm{MTLK}} \varphi$ for $\mathcal{M}, \lambda_\rho^0 \models_{\mathrm{MTLK}} \varphi$. An MTLK formula φ is *existentially valid* in the model $\mathcal{M}$, denoted $\mathcal{M} \models_{\mathrm{MTLK}} \mathbf{E}\varphi$, if, and only if $\mathcal{M}, \lambda_\rho \models_{\mathrm{MTLK}} \varphi$ for some path λ_ρ starting in the initial state of $\mathcal{M}$. Determining whether an MTLK formula φ is existentially valid in a given model is called an *existential model checking problem.*

3 Translation from MTLK to $\mathbf{LTL_qK}$

Abstract model. The set of all the clock valuations is infinite which means that a model has an infinite set of states. We need to abstract the proposed model before we can apply the bounded model checking technique.

Let φ be an MTLK formula and TIS $= (\{L_{\mathbf{i}}, Act_{\mathbf{i}}, \mathcal{X}_{\mathbf{i}}, P_{\mathbf{i}}, \mathcal{V}_{\mathbf{i}}, \mathrm{I}_{\mathbf{i}}, \iota_{\mathbf{i}}\}_{\mathbf{i} \in \mathcal{A} \cup \{\mathcal{E}\}}, \{t_{\mathbf{i}}\}_{\mathbf{i} \in \mathcal{A}}, \{t_{\mathcal{E}}\})$ be a timed interpreted system with $\mathcal{X} = \{x_0, \ldots, x_n\}$. For each $\mathbf{i} \in \mathcal{A} \cup \{\mathcal{E}\}$, let $c_{\mathbf{i}}^{max}$ be the largest constant appearing in intervals of φ and in any enabling condition involving the clock $x_{\mathbf{i}}$ and used in the state invariants and guards of TIS. For two clock valuations v and v' in $\mathbb{N}^{|\mathcal{X}|}$, we say that $v \simeq v'$ iff for each agent $\mathbf{i}$ either $v(x_{\mathbf{i}}) > c_{\mathbf{i}}^{max}$ and $v'(x_{\mathbf{i}}) > c_{\mathbf{i}}^{max}$ or $v(x) \leqslant c_{\mathbf{i}}^{max}$ and $v'(x) \leqslant c_{\mathbf{i}}^{max}$ and $v(x) = v'(x)$.

It is well known, that the relation $\simeq$ is an equivalence relation, what gives rise to construct an finite abstract model. To this end we define the set of possible values of the clock $x_{\mathbf{i}}$ in the abstract model as $\mathbb{D}_{\mathbf{i}} = \{0, \ldots, c_{\mathbf{i}}^{max} + 1\}$. Moreover, for two clock valuations v and v' in $\mathbb{D}_0 \times \ldots \times \mathbb{D}_n \times \mathbb{D}_{\mathcal{E}}$, we say that v' is the *time successor* of v (denoted $succ(v)$) as follows: for each $x \in \mathcal{X}$,

$$succ(v)(x_{\mathbf{i}}) = \begin{cases} v(x_{\mathbf{i}}) + 1, \text{ if } v(x_{\mathbf{i}}) \leqslant c_j^{max}, \\ c_{\mathbf{i}}^{max} + 1, \text{ if } v(x_{\mathbf{i}}) = c_{\mathbf{i}}^{max} + 1. \end{cases}$$

Definition 2. *A tuple* $\widehat{\mathcal{M}} = (\widehat{S}, Act, \widehat{\iota}, \widehat{T}, \widehat{\mathcal{V}})$, *is an* abstract model, *where* $\widehat{\iota} = \prod_{\mathbf{i} \in \mathcal{A} \cup \mathcal{E}} (\iota_{\mathbf{i}} \times \{0\}^{|\mathcal{X}_{\mathbf{i}}|})$ *is the set of all the initial global states,* $\widehat{S} = \prod_{\mathbf{i} \in \mathcal{A} \cup \mathcal{E}} (L_{\mathbf{i}} \times \mathbb{D}_{\mathbf{i}}^{|\mathcal{X}_{\mathbf{i}}|})$ *is the set of all the abstract global states.* $\widehat{\mathcal{V}} : \widehat{S} \to 2^{\mathcal{AP}}$ *is the valuation function such that:* $p \in \widehat{\mathcal{V}}(\widehat{s})$ *iff* $p \in \bigcup_{\mathbf{i} \in \mathcal{A} \cup \mathcal{E}} \widehat{\mathcal{V}}_{\mathbf{i}}(l_{\mathbf{i}}(\widehat{s}))$ *for all* $p \in \mathcal{AP}$*; and* $\widehat{T} \subseteq \widehat{S} \times (Act \cup \tau) \times \widehat{S}$. *Let* $\widetilde{a} \in Act$. *Then, action transition is defined as* $(\widehat{s}, \widetilde{a}, \widehat{s}') \in \widehat{T}$ *iff* $\forall_{\mathbf{i} \in \mathcal{A}} \exists_{\phi_{\mathbf{i}} \in \mathcal{C}(\mathcal{X}_{\mathbf{i}})} \exists_{\mathcal{X}'_{\mathbf{i}} \subseteq \mathcal{X}_{\mathbf{i}}} (t_{\mathbf{i}}(l_{\mathbf{i}}(\widehat{s}), \phi_{\mathbf{i}}, \mathcal{X}'_{\mathbf{i}}, \widetilde{a}) = l_{\mathbf{i}}(\widehat{s}')$ *and* $v_{\mathbf{i}} \models \phi_{\mathbf{i}} \wedge Inv(l_{\mathbf{i}}(\widehat{s}))$ *and*

$v'_{\mathrm{i}}(\widehat{s}') = v_{\mathrm{i}}(\widehat{s})[\mathcal{X}'_{\mathrm{i}} := 0]$ *and* $v'_{\mathrm{i}}(\widehat{s}') \models Inv(l_{\mathrm{i}}(\widehat{s}')))$*; time transition is defined as* $(\widehat{s}, \tau, \widehat{s}') \in \widehat{T}$ *iff* $\forall_{\mathrm{i} \in \mathcal{A} \cup \mathcal{E}}(l_{\mathrm{i}}(\widehat{s}) = l_{\mathrm{i}}(\widehat{s}'))$ *and* $v_{\mathrm{i}}(\widehat{s}) \models Inv(l_{\mathrm{i}}(\widehat{s}))$ *and* $succ(v_{\mathrm{i}}(\widehat{s})) \models Inv(l_{\mathrm{i}}(\widehat{s})))$ *and* $\forall_{\mathrm{i} \in \mathcal{A}}(v'_{\mathrm{i}}(\widehat{s}') = succ(v_{\mathrm{i}}(\widehat{s}')))$ *and* $(v'_{\mathcal{E}}(\widehat{s}') = succ(v_{\mathcal{E}}(\widehat{s})))$.

Definition 3. *A* path *in* $\widehat{\mathcal{M}}$ *is a sequence* $\pi = (s_0, s_1, \ldots)$ *of states such that for each* $j \in \mathbb{N}$*, either* $(s_j \overset{\tau}{\hookrightarrow} s_{j+1})$ *or* $(s_j \overset{\widetilde{a}}{\hookrightarrow} s_{j+1})$ *for some* $\widetilde{a} \in Act$*, and every action transition is preceded by at least one time transition.*

The above definition of the path ensures that the first transition is the time one, and that between each two action transitions at least one time transition appears.

For a path π, $\pi(j)$ denotes the j-th state s_j of π, $\pi[..j] = (\pi(0), \ldots, \pi(j))$ denotes the j-th prefix of π ending with $\pi(j)$, and $\pi^j = (s_j, s_{j+1}, \ldots)$ denotes the j-th suffix of π starting with $\pi(j)$.

Given a path π one can define a function $\zeta_\pi : \mathbb{N} \mapsto \mathbb{N}$ such that for each $j \geqslant 0$, $\zeta_\pi(j)$ is equal to the number of time transitions on the prefix $\pi[..j]$. Let us note that for each $j \geqslant 0$, $\zeta_\pi(j)$ gives the value of the global time in the j-th state of the path π.

The logic $\mathbf{LTL_qK}$. The logic $\mathrm{LTL_q}$ was defined in [8]. $\mathrm{LTL_qK}$ is the fusion of the two underlying languages: $\mathrm{LTL_q}$ and $S5_n$ for the knowledge operators [2].

Let $\mathcal{I}$ be the set of all intervals in $\mathbb{N}$. Let $\mathcal{AP}_{\mathcal{I}} = \{q_{\mathrm{I}} \mid \mathrm{I} \in \mathcal{I}\}$. The $\mathrm{LTL_qK}$ formulae in the negation normal form are defined by the following grammar:

$$\psi ::= \mathbf{true} \mid \mathbf{false} \mid p \mid \neg p \mid q_{\mathrm{I}} \mid \neg q_{\mathrm{I}} \mid \psi \wedge \psi \mid \psi \vee \psi \mid \psi \mathbf{U} \psi \mid \mathbf{G}\psi \mid \mathrm{K_i}\varphi \mid \overline{\mathrm{K}}_{\mathrm{i}}\varphi,$$

where $p \in \mathcal{AP}$ and $q_{\mathrm{I}} \in \mathcal{AP}_{\mathcal{I}}$. The temporal modalities $\mathbf{U}$ and $\mathbf{G}$ are, respectively, named as the *until* and the *always*. The derived basic temporal modality for *eventually* is defined in the standard way: $\mathbf{F}\psi \overset{def}{=} \mathbf{true}\mathbf{U}\psi$.

In order to improve readability, in the following definition we write $\langle \pi, m \rangle \models_k \psi$ instead of $\widehat{\mathcal{M}}, \langle \pi, m \rangle \models_k \psi$, for any $\mathrm{LTL_qK}$ formula ψ.

Definition 4. *The* satisfiability *relation* $\models^d$*, which indicates truth of an* $\mathrm{LTL_qK}$ *formula in the abstract model* $\widehat{\mathcal{M}}$ *along the path* π *with the starting point* m *and at the depth* $d \geqslant m$*, is defined inductively as follows:*

- $\langle \pi, m \rangle \models^d \mathbf{true}$, $\quad \langle \pi, m \rangle \not\models^d \mathbf{false}$,
- $\langle \pi, m \rangle \models^d p$ iff $p \in \mathcal{V}(\pi(d))$, $\quad \langle \pi, m \rangle \models^d \neg p$ iff $p \notin \mathcal{V}(\pi(d))$,
- $\langle \pi, m \rangle \models^d q_{\mathrm{I}}$ iff $\zeta_\pi(d) - \zeta_\pi(m) \in \mathrm{I}$, $\quad \langle \pi, m \rangle \models^d \neg q_{\mathrm{I}}$ iff $\zeta_\pi(d) - \zeta_\pi(m) \notin \mathrm{I}$,
- $\langle \pi, m \rangle \models^d \alpha \wedge \beta$ iff $\langle \pi, m \rangle \models^d \alpha$ and $\langle \pi, m \rangle \models^d \beta$,
- $\langle \pi, m \rangle \models^d \alpha \vee \beta$ iff $\langle \pi, m \rangle \models^d \alpha$ or $\langle \pi, m \rangle \models^d \beta$,
- $\langle \pi, m \rangle \models^d \alpha \mathbf{U} \beta$ iff $(\exists j \geqslant d)(\langle \pi, d \rangle \models^j \beta$ and $(\forall d \leqslant i < j)\, \langle \pi, d \rangle \models^i \alpha)$,
- $\langle \pi, m \rangle \models^d \mathbf{G}\beta$ iff $(\forall j \geqslant d)\, \langle \pi, d \rangle \models^j \beta$,
- $\langle \pi, m \rangle \models^d \mathrm{K_i}\alpha$ iff $(\forall \pi' \in \Pi)(\forall j \geq d)(\langle \pi', j \rangle \sim_{\mathrm{i}} \langle \pi, d \rangle$ implies $\langle \pi', d \rangle \models^j \alpha)$,
- $\langle \pi, m \rangle \models^d \overline{\mathrm{K}}_{\mathrm{i}}\alpha$ iff $(\exists \pi' \in \Pi)(\exists j \geq d)(\langle \pi', j \rangle \sim_{\mathrm{i}} \langle \pi, d \rangle$ and $'\, \langle \pi', d \rangle \models^j \alpha)$.

An $\mathrm{LTL_qK}$ formula ψ *existentially holds* in the model $\widehat{\mathcal{M}}$, written $\widehat{\mathcal{M}} \models \mathbf{E}\psi$, if, and only if $\widehat{\mathcal{M}}, \langle\pi, 0\rangle \models^0 \psi$ for some path π starting at the initial state. The *existential model checking problem* asks whether $\widehat{\mathcal{M}} \models \mathbf{E}\psi$.

Translation. The translation from MTLK to $\mathrm{LTL_qK}$ is based on translation presented in [8]. Let $p \in \mathcal{AP}$, α and β be formulae of MTLK. We define the translation from MTLK into $\mathrm{LTL_qK}$ as a function $\mathrm{tr} : \mathrm{MTLK} \to \mathrm{LTL_qK}$ in the following way:

- $\mathrm{tr}(\mathbf{true}) = \mathbf{true}$, $\mathrm{tr}(\mathbf{false}) = \mathbf{false}$, $\mathrm{tr}(p) = p$, $\mathrm{tr}(\neg p) = \neg p$,
- $\mathrm{tr}(\alpha \wedge \beta) = \mathrm{tr}(\alpha) \wedge \mathrm{tr}(\beta)$, $\mathrm{tr}(\alpha \vee \beta) = \mathrm{tr}(\alpha) \vee \mathrm{tr}(\beta)$,
- $\mathrm{tr}(\alpha \mathbf{U}_{\mathrm{I}} \beta) = \mathrm{tr}(\alpha)\mathbf{U}(q_{\mathrm{I}} \wedge \mathrm{tr}(\beta))$, $\mathrm{tr}(\mathbf{G}_{\mathrm{I}}\beta) = \mathbf{G}(\neg q_{\mathrm{I}} \vee \mathrm{tr}(\beta))$
- $\mathrm{tr}(\mathrm{K}\alpha) = \mathrm{K}\alpha$, $\mathrm{tr}(\overline{\mathrm{K}}\alpha) = \overline{\mathrm{K}}\alpha$.

Observe that the translation of literals as well as logical connectives and epistemic operators is straightforward. The translation of the $\mathbf{U}_{\mathrm{I}}$ operator ensures that β holds somewhere in the interval I (this is expressed by the requirement $q_{\mathrm{I}} \wedge \mathrm{tr}(\beta)$), and α holds always before β. Similarly, the translation of the $\mathbf{G}_{\mathrm{I}}$ operator ensures that β always holds in the interval I (this is expressed by the requirement $\neg q_{\mathrm{I}} \vee \mathrm{tr}(\beta)$).

Theorem 1. *Let* TIS *be a timed interpreted system,* φ *an* MTLK *formula, and* $\mathcal{M}$ *the model for the timed interpreted system* TIS*, and* $\widehat{\mathcal{M}}$ *the abstract model for the timed interpreted system* TIS *and the formula* φ*. Then,* $\mathcal{M} \models \mathbf{E}\varphi$ *if, and only if* $\widehat{\mathcal{M}} \models \mathbf{E}\mathrm{tr}(\varphi)$.

4 Bounded Model Checking

In this section we define a *bounded semantics* for $\mathrm{LTL}_q\mathrm{K}$ in order to translate the *existential model checking problem* for $\mathrm{LTL}_q\mathrm{K}$ into the satisfiability problem.

Bounded semantics. To define the bounded semantics one needs to represent infinite paths in a model in a special way. To this aim, we recall the notions of *k-paths* and *loops* [7].

Definition 5. *Let* $\widehat{\mathcal{M}}$ *be the abstract model,* $k \in \mathbb{N}$*, and* $0 \leqslant l \leqslant k$*. A* k-path *is a pair* (π, l)*, also denoted by* π_l*, where* π *is a finite sequence* $\pi = (s_0, \ldots, s_k)$ *of states such that for each* $0 \leqslant j < k$*, either* $(s_j \overset{\tau}{\hookrightarrow} s_{j+1})$ *or* $(s_j \overset{\widetilde{a}}{\hookrightarrow} s_{j+1})$ *for some* $\widetilde{a} \in Act$*, and every action transition is preceded by at least one time transition. A* k*-path* π_l *is a* loop*, written* $\circlearrowleft(\pi_l)$ *for short, if* $l < k$ *and* $\pi(k) = \pi(l)$.

If a k-path π_l is a loop it represents the infinite path of the form uv^ω, where $u = (\pi(0), \ldots, \pi(l))$ and $v = (\pi(l+1), \ldots, \pi(k))$. We denote this unique path by $\widetilde{\pi}_l$. Note that for each $j \in \mathbb{N}$, $\widetilde{\pi}_l^{\,l+j} = \widetilde{\pi}_l^{\,k+j}$.

In the definition of bounded semantics for variables from $\mathcal{AP}_{\mathcal{I}}$ one needs to use only a finite prefix of the sequence $(\zeta_{\widetilde{\pi}_l}(0), \zeta_{\widetilde{\pi}_l}(1), \ldots)$. Namely, for a k-path

π_l that is not a loop the prefix of the length k is needed, and for a k-path π_l that is a loop the prefix of the length $k+k-l$ is needed.

In order to improve readability, in the following definition we write $\langle \pi_l, m\rangle \models_k \psi$ instead of $\widehat{\mathcal{M}}, \langle \pi_l, m\rangle \models_k \psi$, for any $\mathrm{LTL}_q\mathrm{K}$ formula ψ.

Definition 6 (Bounded semantics). *Let $\widehat{\mathcal{M}}$ be the abstract model, π_l be a k-path in $\widehat{\mathcal{M}}$, and $0 \leqslant m, d \leqslant k$. The relation $\models_k^d$ is defined inductively as follows:*

- $\langle \pi_l, m\rangle \models_k^d \textbf{true}$, $\quad \langle \pi_l, m\rangle \not\models_k^d \textbf{false}$,
- $\langle \pi_l, m\rangle \models_k^d p$ *iff* $p \in \mathcal{V}(\pi_l(d))$, $\quad \langle \pi_l, m\rangle \models_k^d \neg p$ *iff* $p \notin \mathcal{V}(\pi_l(d))$,
- $\langle \pi_l, m\rangle \models_k^d q_{\mathrm{I}}$ *iff* $\begin{cases} \zeta_{\widetilde{\pi}_l}(d) - \zeta_{\widetilde{\pi}_l}(m) \in \mathrm{I}, & \text{if } \pi_l \text{ is not a loop},\\ \zeta_{\widetilde{\pi}_l}(d) - \zeta_{\widetilde{\pi}_l}(m) \in \mathrm{I}, & \text{if } \pi_l \text{ is a loop and } d \geqslant m,\\ \zeta_{\widetilde{\pi}_l}(d+k-l) - \zeta_{\widetilde{\pi}_l}(m) \in \mathrm{I}, & \text{if } \pi_l \text{ is a loop and } d < m, \end{cases}$
- $\langle \pi_l, m\rangle \models_k^d \neg q_{\mathrm{I}}$ *iff* $\langle \pi_l, m\rangle \not\models_k^d q_{\mathrm{I}}$
- $\langle \pi_l, m\rangle \models_k^d \alpha \wedge \beta$ *iff* $\langle \pi_l, m\rangle \models_k^d \alpha$ *and* $\langle \pi_l, m\rangle \models_k^d \beta$,
- $\langle \pi_l, m\rangle \models_k^d \alpha \vee \beta$ *iff* $\langle \pi_l, m\rangle \models_k^d \alpha$ *or* $\langle \pi_l, m\rangle \models_k^d \beta$,
- $\langle \pi_l, m\rangle \models_k^d \alpha \mathbf{U} \beta$ *iff* $(\exists_{d \leqslant j \leqslant k})\big(\langle \pi_l, d\rangle \models_k^j \beta$ *and* $(\forall_{d \leqslant i < j})\, \langle \pi_l, d\rangle \models_k^j \alpha\big)$
  or $\big(\circlearrowleft(\pi_l)$ *and* $(\exists_{l<j<d})\, \langle \pi_l, d\rangle \models_k^j \beta$ *and* $(\forall_{l<i<k})\ \langle \pi_l, d\rangle \models_k^j \alpha$
  and $(\forall_{d \leqslant i \leqslant k})\ \langle \pi_l, d\rangle \models_k^j \alpha\big)$,
- $\langle \pi_l, m\rangle \models_k^d \mathbf{G}\beta$ *iff* $\circlearrowleft(\pi_l)$ *and* $(\forall_{j \leqslant k}) j \geqslant \min(d, l)$ *implies* $\langle \pi_l, d\rangle \models_k^j \beta$,
- $\langle \pi_l, m\rangle \models_k^d \overline{\mathrm{K}}_{\mathbf{i}}\alpha$ *iff* $(\exists \pi'_{l'} \in \Pi_k)(\exists d \leq j \leq k)(\langle \pi_l, d\rangle \models_k^j \alpha$ *and* $\pi(d) \sim_{\mathbf{i}} \pi'(j))$.

An $\mathrm{LTL}_q\mathrm{K}$ formula ψ *existentially k-holds* in the model $\widehat{\mathcal{M}}$, written $\widehat{\mathcal{M}} \models_k \mathbf{E}\psi$, if, and only if $\widehat{\mathcal{M}}, \langle \pi, 0\rangle \models_k^0 \psi$ for some path π starting at the initial state.

Theorem 2. *Let* TIS *be a timed interpreted system, φ an MTLK formula, and $\widehat{\mathcal{M}}$ the abstract model for the timed interpreted system* TIS *and the formula φ. Moreover, let $\psi = \mathrm{tr}(\varphi)$. Then, $\widehat{\mathcal{M}} \models \mathbf{E}\psi$ if, and only if there exists $k \geqslant 0$ such that $\widehat{\mathcal{M}} \models_k \mathbf{E}\psi$.*

Translation to SAT. The last step of our method is the standard one (see [4,7]). It consists in encoding the transition relation of $\widehat{\mathcal{M}}$ and the $\mathrm{LTL_qK}$ formula $\mathrm{tr}(\varphi)$. The only novelty lies in encoding of the finite prefix of the sequence $(\zeta_{\widetilde{\pi}_l}(0), \zeta_{\widetilde{\pi}_l}(1), \ldots)$. The translation to SAT results in the propositional formula $[\widehat{\mathcal{M}}, \mathrm{tr}(\varphi)]_k$ with the property expressed in the following theorem.

Theorem 3. *Let $\widehat{\mathcal{M}}$ be an abstract model. Then, for every $k \in \mathbb{N}$, $\widehat{\mathcal{M}} \models_k^d$ $\mathbf{E}\mathrm{tr}(\varphi)$ if, and only if, the propositional formula $[\widehat{\mathcal{M}}, \mathrm{tr}(\varphi)]_k$ is satisfiable.*

5 Experimental Results

In this section we experimentally evaluate the performance of our new translation described in the previous sections. We have conducted the experiments using Timed Generic Pipeline Paradigm (TGPP) [8].

The Timed Generic Pipeline Paradigm (TGPP) TIS model shown in Fig. 1 consists of Producer producing data within a certain time interval ($[a, b]$) or being inactive, Consumer receiving data within a certain time interval ($[c, d]$) or being inactive within a certain time interval ($[g, h]$), and a chain of n intermediate Nodes which can be ready for receiving data within a certain time interval ($[c, d]$), processing data within a certain time interval ($[e, f]$) or sending data. We assume that $a = c = e = g = 1$ and $b = d = f = h = 2 \cdot n + 2$, where n represents number of nodes in the TGPP.

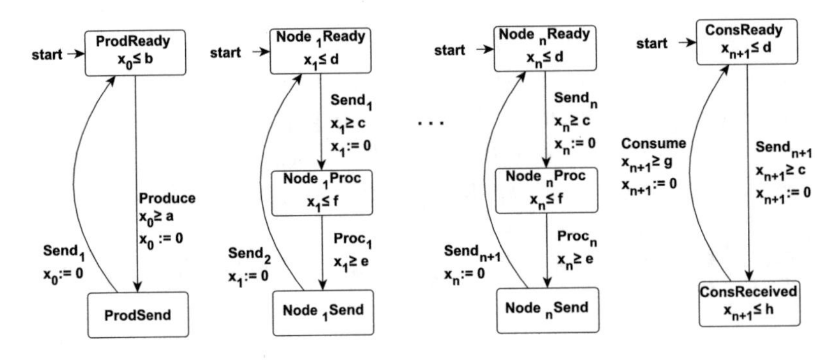

Fig. 1. The TGPP system.

We have tested the TGPP timed interpreted system model, scaled in the number of intermediate nodes on the following MTLK formulae that existentially hold in the model of TGPP (n is the number of nodes):

- $\varphi_1 = \mathbf{G}(\mathrm{K}_P(\mathit{ProdSend} \Rightarrow \mathbf{F}_{[0,2n+2)}(\mathit{ConsReceived})))$. It expresses that Producer knows that each time Producer produces data, then Consumer receives this data not later than in $2n + 1$ time units.
- $\varphi_2 = \mathrm{K}_P(\mathbf{F}_{[0,2n+1)}(\mathit{ConsReceived}))$. It states that Producer knows that eventually Consumer will receive data not later than in $2n + 1$ time units.
- $\varphi_3 = \mathrm{K}_C(\mathrm{K}_P(\mathbf{F}_{[0,2n+2)}(\mathit{ConsReceived})))$. It expresses that Consumer knows that Producer knows that eventually Consumer will receive data not later than in $2n + 2$ time units.
- $\varphi_4 = \mathrm{K}_P(\mathit{ConsReceived} \Rightarrow \mathbf{F}_{[0,2n+1)}(\neg \mathit{ConsReceived}))$. It states that Producer knows that time Consumer receives data, then Consumer is ready to receive data no later than $2n + 1$ time units after that Consumer will receive data.

Performance evaluation. We have performed our experimental results on a computer equipped with I7-3770 processor, 32 GB of RAM, and the operating system Linux. Our SAT-based BMC algorithms are implemented as standalone programs written in the programming language C++. We used the state of the art SAT-solver CryptoMiniSAT.

All the benchmarks together with instructions on how to reproduce our experimental results can be found at the web page tinyurl.com/sat-bmc-tis-emtlk.

The number of considered k-paths for the properties φ_1, φ_2, and φ_4, is equal to 2, and for the property φ_3 is equal to 3.

As one can see from the line charts in Fig. 2 showing the total time and the memory consumption for all the tested properties, the experimental results confirm that our new SAT-based BMC for TIS and for EMTLK is indeed feasible. Moreover, we can observe that as in the case of other known SAT-based BMC methods, our new approach is also sensitive on the size of the counterexample. The size of the resulting SAT formula, measured by the number of clauses, grows with the number of components and depends also on the structure of the input MTLK formula.

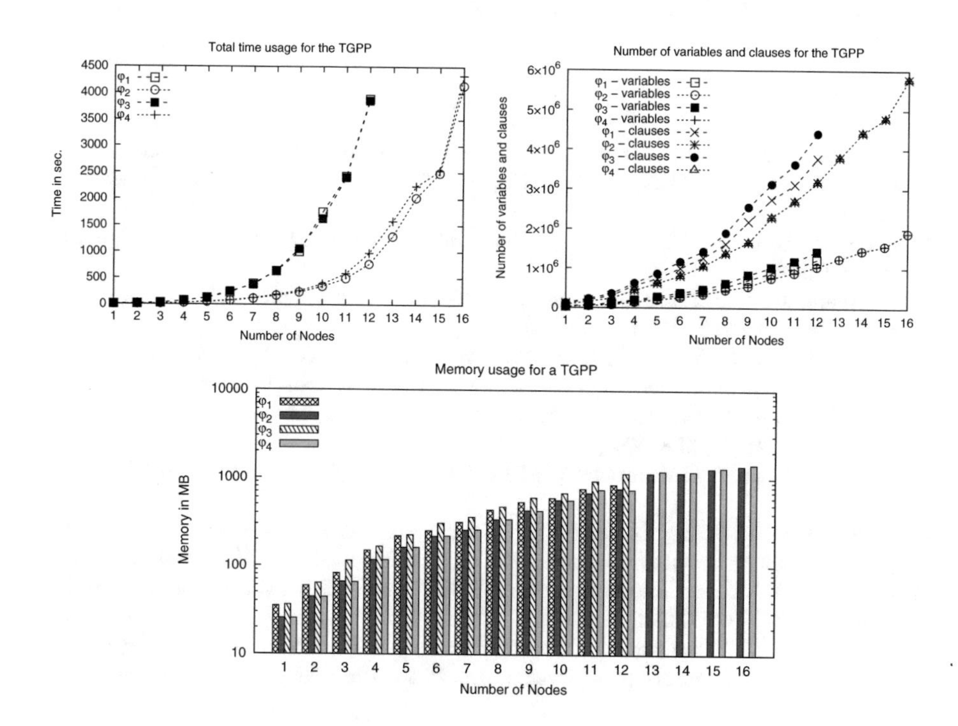

Fig. 2. TGPP with n nodes.

6 Conclusions

We have proposed, implemented, and experimentally evaluated SAT-based BMC method for timed interpreted systems and for properties expressible in MTLK with the semantics over timed interpreted systems. The method is based on a translation of the existential model checking for MTLK to the existential model checking for $\mathrm{LTL_qK}$, and then on the translation of the existential model checking for $\mathrm{LTL_qK}$ to the propositional satisfiability problem.

We believe that our approach is much better than the approach based on translation to HLTLK as we could see in [8]. The reason is that the new method needs only one new path for temporal operators, does not need any new clocks, and does not need any new transitions. The paper presents preliminary experimental results only, but they show that the proposed verification method is quite efficient and worth exploring.

Therefore, in our future work we are going to define the SMT-based BMC encoding for TIS and for $\mathrm{LTL_qK}$ and compare this encoding with the SAT-based encoding, and we would like to compare our new implementation with the implementation of the translation from MTLK to HLTLK presented in [6].

References

1. Biere, A., Cimatti, A., Clarke, E., Zhu, Y.: Symbolic model checking without BDDs. In: Proceedings of TACAS 1999. LNCS, vol. 1579, pp. 193–207. Springer, Heidelberg (1999)
2. Fagin, R., Halpern, J.Y., Moses, Y., Vardi, M.Y.: Reasoning About Knowledge. MIT Press, Cambridge (1995)
3. Koymans, R.: Specifying real-time properties with metric temporal logic. Real-Time Syst. **2**(4), 255–299 (1990)
4. Męski, A., Penczek, W., Szreter, M., Woźna-Szcześniak, B., Zbrzezny, A.: Bdd-versus sat-based bounded model checking for the existential fragment of linear temporal logic with knowledge: algorithms and their performance. Auton. Agent. Multi-Agent Syst. **28**(4), 558–604 (2014)
5. Wooldridge, M.: An Introduction to Multi-agent Systems, 2nd edn. Wiley, New York (2009)
6. Woźna-Szcześniak, B., Zbrzezny, A.: Checking EMTLK properties of timed interpreted systems via bounded model checking. Studia Logica **104**(4), 641–678 (2016)
7. Zbrzezny, A.: A new translation from ECTL* to SAT. Fundamenta Informaticae **120**(3–4), 377–397 (2012)
8. Zbrzezny, A.M., Zbrzezny, A.: Simple bounded MTL model checking for discrete timed automata (extended abstract). In: Proceedings of CS&P 2016, pp. 37–48 (2016)

An Algorithm for Allocating Structured Tasks in Multi-Robot Scenarios

Tulio L. Basegio(✉) and Rafael H. Bordini

Postgraduate Programme in Computer Science, School of Informatics (FACIN),
Pontifical Catholic University of Rio Grande do Sul (PUCRS),
Porto Alegre, RS, Brazil
tulio.basegio@acad.pucrs.br, rafael.bordini@pucrs.br

Abstract. Task allocation is an important aspect in dealing with coordination problems. However, there are challenges in developing appropriate strategies for multi-robot teams in such a way that robots perform their operations efficiently. Real-world scenarios usually require the use of heterogeneous robots and execution of tasks with different structures and constraints. In this paper we propose a dynamic, decentralised task allocation mechanism considering different types of tasks for heterogeneous robot teams playing different roles and carrying out tasks according to their own capabilities. We have run several simulations in order to evaluate the proposed mechanism. The results indicate that the proposed mechanism scales well and provides near-optimal allocations.

Keywords: Task allocation · Multi-robot systems · Multi-agent systems

1 Introduction

One of the challenges in developing multi-robot systems today is the design of coordination strategies in such a way that robots perform their operations efficiently [11]. Without such strategies, the use of multi-robot systems in complex scenarios becomes limited or even unfeasible.

An important aspect considered in coordination problems is task allocation [5, 11]. There are several features that should be considered by a mechanism for allocating tasks to multiple robots in real-world scenarios such as considering the heterogeneity of robots, the impact of individual variability to assign specific roles to individual robots, and the definition and allocation of different types of tasks. This is particularly true for disasters such as flooding [8].

During a rescue phase in a flooding disaster, teams are called into action to work in tasks such as locating and rescuing victims [6]. Such teams are normally organised by a hierarchy model [7], with individuals playing different roles during a mission. The execution of tasks during the rescue stage poses a number of risks to the teams. Using robots in a coordinated way to help the team may minimise those risks. It was the typical tasks in flooding rescue that inspired us to work on

G. Jezic et al. (eds.), *Agent and Multi-Agent Systems: Technology and Applications*, Smart Innovation, Systems and Technologies 74,
DOI 10.1007/978-3-319-59394-4_10

a task allocation mechanism that could address different types of tasks. Although the actual flooding rescue tasks we are dealing with is not the focus of this paper, the algorithm we present here was inspired and is being developed for such an application in a multi-institution project[1] to address disaster response, particularly in case of flooding.

Although real-world scenarios typically require the use of heterogeneous robots and execution of tasks with different constraints and structures, most of the solutions found in literature either deal with the allocation of one type of task only (mainly atomic tasks) or focus on one type of constraint only or have high computational cost. The main contribution of our work is a decentralised mechanism for the dynamic allocation of different types of tasks to heterogeneous robot teams, considering that they can play different roles and carry out tasks according to their own capabilities. The proposed mechanism extends the work presented in [5], adding to that approach different types of tasks, the use of roles, and verification of constraints related to the heterogeneity of robots. In this paper we use the term agent to refer to the main control software of an individual robot, so our multi-robot system is effectively treated as a multi-agent system.

2 Multi-Robot Task Allocation (MRTA)

Task allocation among multiple robots (and more generally among multiple agents) consists of identifying which robots should perform which tasks in order to achieve cooperatively the global objective function in the best possible way. When a task is executed by a robot, it contributes some value to the robots' objective and, consequently, to the global team goals. This value is referred as utility, and can combine several factors (such as payoff to be received, the costs incurred, etc.) [3]. In this paper we assume that each robot is able to calculate its utility for each task.

Different types of tasks can be used to address different problems involved in real-world scenarios, which cannot be adequately represented by only one type of task due to their complex structures and other domain-specific characteristics. In this paper we use the following type of tasks defined in [12]:

- Atomic task (AT): a task is atomic if it cannot be decomposed into subtasks.
- Decomposable simple task (DS): a task that can be decomposed into a set of atomic subtasks or other decomposable simple tasks as long as the different decomposed parts have to be carried out by the same robot.
- Compound task (CT): task that can be decomposed into a set of atomic or compound subtasks. When each of the subtasks need to be allocated to a different robot we call it CN task (N subtasks that need exactly N robots). When there is no constraints, the subtasks can be allocated to one up to M robots, where M is the number of subtasks (CM tasks).

[1] Disaster Robotics "Pro-Alertas" (funded by CAPES – Pro-Alertas) https://disaster-robotics-proalertas.github.io.

2.1 Problem Statement

Here we formally state our multi-robot task allocation as an optimisation problem. Consider that there are n_r available robots $R = \{r_1, \ldots, r_{n_r}\}$. Consider also that there are n_t tasks $T = \{t_1, \ldots, t_{n_t}\}$ and n_{st} subtasks $ST = \{st_1, \ldots, st_{n_{st}}\}$ where each subtask st_k belongs to exactly one task t_j. Each task t_j has one or more subtasks from ST, and we use N_j for the specific number of subtasks that the jth task has. Consider further that there are n_q types of tasks $Q = \{q_1, \ldots, q_{n_q}\}$ and that each task t_j is associated to exactly one type of task from Q. Each type of task $q \in Q$ has a minimum and maximum (min_q, max_q) number of subtasks that a robot must take on when allocating to itself subtasks of a single task of type q. We use min_j and max_j for the minimum and maximum number of subtasks that a robot must take when allocating subtasks from a task t_j. Note that with the min and max constraints we can represent tasks such as DS, CN and CM tasks. For example, a robot trying to allocate a DS task must take all of the N_j subtasks, i.e. it must take a minimum of N_j and a maximum of N_j subtasks since the type DS requires the allocation of all subtasks to *exactly one* robot. Similarly, a robot trying to allocate a CN task with N_j subtasks must take only *one* subtask, i.e. it must take a minimum and a maximum of *one* subtask. A robot trying to allocate a CM task with N_j subtasks must take a minimum of *one* and a maximum of N_j subtasks. Consider also that there are n_c capabilities $C = \{c_1, \ldots, c_{n_c}\}$ and n_e roles $E = \{e_1, \ldots, e_{n_e}\}$. Each role e is associated with the capabilities a robot must have in order to be able to play that role. Each robot r_i has a set of capabilities, which determine the set of roles it is able to play. Each subtask st_k is associated with a set of roles a robot must be able to play in order to execute it. Finally, we assume that a task $t_j \in T$ is considered allocated if all of its subtasks were allocated to robots following the constraints described above. Each subtask may be allocated to at most one robot, and each robot r_i can perform at most L_i subtasks (i.e., its task limit). Let us consider the following binary variables: f_{ik} indicates whether r_i is assigned to st_k; p_{jk} indicates whether st_k belongs to t_j; g_{yx} indicates whether e_y requires c_x; v_{ix} indicates whether r_i has c_x; h_{ky} indicates whether st_k requires e_y and; z_{iy} indicates whether r_i is able to play e_y. Finally, let $u_{ik} \in \mathbb{R}$ be the utility of the allocation of r_i to st_k.

The objective of our MRTA problem is to find an allocation that maximises the sum of utilities while satisfying all the above constraints, and can be stated as follows:

Objective:

$$\max_{\{f_{ik}\}} \sum_{i=1}^{n_r} \sum_{k=1}^{n_{st}} u_{ik} \cdot f_{ik} \tag{1}$$

subject to:

$$\sum_{i=1}^{n_r} f_{ik} \leq 1 \;\; \forall k = 1, \ldots, n_{st}. \tag{2}$$

$$\sum_{k=1}^{n_{st}} f_{ik} \leq L_i \;\; \forall i = 1, \ldots, n_r. \tag{3}$$

$$\sum_{j=1}^{n_t} p_{jk} = 1 \;\; \forall k = 1, ..., n_{st}. \tag{4}$$

$$\sum_{x=1}^{n_c} g_{yx} \leq n_c \;\; \forall y = 1, ..., n_e. \tag{5}$$

$$\sum_{x=1}^{n_c} v_{ix} \leq n_c \;\; \forall i = 1, ..., n_r. \tag{6}$$

$$\sum_{y=1}^{n_e} z_{iy} \leq n_e \;\; \forall i = 1, ..., n_r. \tag{7}$$

$$\sum_{y=1}^{n_e} h_{ky} \leq n_e \;\; \forall k = 1, ..., n_{st}. \tag{8}$$

$$\sum_{k=1}^{n_{st}} f_{ik}.p_{jk} \geq min_j \quad \forall j \text{ s.t. } t_j \in T; \; i = 1, \ldots, n_r. \tag{9}$$

$$\sum_{k=1}^{n_{st}} f_{ik}.p_{jk} \leq max_j \quad \forall j \text{ s.t. } t_j \in T; \; i = 1, \ldots, n_r. \tag{10}$$

$$\sum_{x=1}^{n_c} g_{yx}.v_{ix}.z_{iy} = \sum_{x=1}^{n_c} g_{yx}.z_{iy} \quad \forall i = 1, ..., n_r; \; y = 1, \ldots, n_e. \tag{11}$$

$$\sum_{y=1}^{n_e} h_{ky}.z_{iy}.f_{ik} = \sum_{y=1}^{n_e} h_{ky}.f_{ik} \quad \forall i = 1, ..., n_r; \; k = 1, \ldots, n_{st}. \tag{12}$$

Equations (9) and (10) state, for each task, the constraints on the number of subtasks a robot must allocate to itself based on the type of that task.

3 The Proposed Approach

We consider the existence of an **organisation** that is responsible for announcing tasks that need to be carried out by the agents. The tasks are published in a **blackboard** that can be viewed by all the available **agents**. Finally, the **environment** is the place where agents carry out the tasks. By identifying new tasks, the agents begin the allocation process based on the proposed mechanism.

3.1 The Proposed Mechanism

The proposed mechanism allow us to work with types of tasks such as decomposable simple tasks (DS) and compound tasks (CN and CM) through the definition of the minimum and maximum number of subtasks a robot must take from tasks of each type. Atomic tasks are considered as subtasks of both compound and decomposable simple tasks. The tasks require agents able to play particular roles to carry them out. We assume that an agent can have different capabilities and also a maximum number of tasks that can be allocated to itself.

Our mechanism is composed by algorithms that are executed by each agent, characterising a decentralised solution. The initial algorithm is Algorithm 1 which receives as input: the list of tasks to be carried out by the agents as currently available in the blackboard and; the list of roles within the organisation. The first step for an agent is to select only the tasks that are compatible with the roles it can play (line 3 calls the Algorithm 2 responsible for this). Knowing the tasks that the agent can perform, it calls getMinMaxTaskType function (line 4) which updates for each task the minimum and maximum number of subtasks that a robot must take from each task. Finally, it calls Algorithm 3 (line 5) which starts the allocation process for the tasks that can possibly be allocated to that agent. Algorithm 2 receives as input the list of all blackboard tasks that need to be carried out as well as the description of all the roles (with their required capabilities) defined within the organisation so that the agent can identify the roles it can play. Initially, the algorithm identifies the possible roles the agent may play by considering its capabilities and the capabilities required for each role (lines 1 to 12). Then, given the roles the agent is able to play, the algorithm identifies which tasks can be allocated to the agent (13 to 17).

Algorithm 1. `startAllocation`(*blackboardTasks*,*organisationRoles*)

```
1: allocatedSubtasks = ∅;
2: candidates = ∅;
3: possibleTasks ← getPossibleTasks(blackboardTasks, organisationRoles);
4: possibleTasks ← getMinMaxTaskType(possibleTasks)
5: taskAllocation(possibleTasks, candidates, allocatedSubtasks);
```

Algorithm 2. `getPossibleTasks`(*blackboardTaskList*, *rolesList*)

```
    Let agentCapabilities be the list of agent's capabilities;
2:  for all role r_k in rolesList do
      validRole ← true
4:    for all capabilityRequired in r_k do
        if capabilityRequired not in agentCapabilities then
6:        validRole ← false
        end if
8:    end for
      if validRole = true then
10:     agentRoles.add(r_k)
      end if
12: end for
    for all task t_j in blackboardTaskList do
14:   if t_j.role in agentRoles then
        possibleTasks.add(t_j)
16:   end if
    end for
18: return possibleTasks
```

Algorithm 3 receives as input the list of possible tasks to be allocated to agent r_i. Initially, the algorithm checks if the number of allocated tasks (na_i) so far is lower than agent capacity (the task limit l_i) and, if so, it begins the analysis of the tasks in order to identify those that could possibly be allocated (lines 5 to 16). The analysis goes through each task as described below. We use l'_i to refer to the difference between l_i and na_i. For each task t_j, the algorithm identifies the number of subtasks the agent can select as candidates to allocation (line 12). In order to get this value the algorithm first checks if the agent has capacity to select the minimum number of subtasks required by task t_j (line 10) and if there is still subtasks of task t_j that are not in the list of allocated subtasks (line 11). Then, it selects as candidates the *nToAlloc* best subtasks from task t_j (line 13). The choice of candidates is carried out based on the utility of each subtask.

From the subtasks selected as candidate, the algorithm selects the best subtasks for allocation considering the task limit for the agent (line 17). For the selected subtasks, it then calculates the bids to send to the other agents. The current calculation is based on the formula (see Eq. 16) introduced in [5], where a new bid for a subtask is based on the old bid plus the utility that will be lost if the robot has to be assigned to the next best subtask instead of the current one. Lines 19 to 22 refer to the calculation of the bids and their broadcasting.

When an agent receives a bid from other agent that is greater than the bid the agent itself provided when allocating the subtask to itself, the agent has to remove the subtask from its allocated subtasks and then it tries to allocate further subtasks by executing Algorithm 3. The process is repeated until the

Algorithm 3. `taskAllocation`(*possibleTasks*,*candidates*, *allocatedSubtasks*)

```
    Let l be the agent's max. number of allowed concurrent tasks;
    na ← allocatedSubtasks.size;
    l' ← l − na;
 4: if l' > 0 then
      for all task t_j in possibleTasks do
        minT ← t_j.minSubTask;
        maxT ← t_j.maxSubTask;
 8:     naT ← allocatedSubtasks.countSubtasksFrom(t_j);
        notAllocT ← maxT − naT;
        if l' >= minT then
          if maxT > naT then
12:         nToAlloc ← min[l', maxT, notAllocT];
            candidates.add(nBestSubtasks(t_j.subTasks, nToAlloc));
          end if
        end if
16:   end for
      bestCandidates ← getBest(candidates, l');
      allocatedSubtasks.add(bestCandidates);
      for all subtask in bestCandidates do
20:     calculateBidValue(subtask);
        communicate new bid values to all other agents;
      end for
    end if
```

agents agree on the allocation, that is, until the self-allocated subtasks do not undergo any further modifications.

3.2 Coping with Partially Allocated Tasks

When the number of subtasks is greater than the total capacity of the agents, the final allocation can result in tasks not completely allocated, that is, tasks in which at least one subtask was not allocated to any agent. As it makes sense that tasks must be completely allocated or not allocated at all, agents make extra effort to identify those tasks and to try reallocate them. Thus, at the end of the allocation previously described the agents identify the tasks partially allocated and remove any of their subtasks from their allocation list. After this step, there may be several completely unallocated tasks and also agents which have space for new allocations. The allocation of the remaining tasks occurs by running an simple auction to fully allocate one task by time. The order in which the tasks will be auctioned is relevant since the preference order may be different for the agents. We use a social-choice algorithm based on voting (Borda count) to achieve such an agreement on that ordering. The auctions are distributed and there is no an auctioneer role.

4 Evaluation

This section compares the performance of the proposed mechanism with the optimal solution using Monte-Carlo simulations. Our mechanism was implemented using a framework for multi-agent systems development called JaCaMo [1]. The GLPK (GNU Linear Programming Kit [13]) solver was used to obtain (centralised) optimal solutions for comparison with our results.

By performance we mean how close our results are to the optimal solution (100%). The coefficient of variation (Standard Deviation/Mean) was used as a measure of dispersion that describes the amount of variability relative to the mean. The number of bids sent is also measured to assess the impact of the different settings on the network traffic.

The simulations were run by varying a different parameter at each setting (Table 1). In all simulations the subtasks of different types of task (CN, CM, DS) were uniformly distributed and for each robot we randomly selected the utility values to each subtask based on the utility range. In the settings 1 and 2 were run simulations with the total capacity of the agents greater than or equal to the total number of subtasks. The results for each variation in these simulations were averaged over one-hundred iterations each. For the setting 3 the simulations were run with more subtasks than the total capacity of the agents. The average results were calculated from twenty iterations for each variation.

Setting 1 – Varying the Number of Agents: First, simulations were run considering agents with capabilities to play any role and thus are able to carry out any task. Figure 1a shows that the performance of the proposed solution

Table 1. Setting of variations used in the simulations.

Setting	Agents	Subtasks	Limit	Utility range
1	5, 10, 15, 20, 25, 30, 35, 40	24	5	1–6
2	10	15, 30, 45, 60	7	1–15
3	3	21, 28, 35	6	1–6

improves and is closer to the optimal solution as we increase the number of agents. Also, the coefficient of variation between the solutions is lower for larger teams (Fig. 1b). Regarding the number of bids, Fig. 1c shows that the average number of bids by agent remains stable for larger number of agents while the average number of bids by task increases since there are more agents bidding the same tasks. Then we run simulations by randomly assigning different capabilities to each robot (from 1 up to 4 capabilities). Thus, some robots may not be able to play some roles, limiting the tasks they are able to carry out. Figure 2 shows that our approach is able to maintain reasonable results, similar to those of the first simulation, even when the capabilities of the robots limit the tasks that they can perform. The average execution time in all these simulations was 4 s for each simulation with 5 agents, and 15 s with 35 agents.

Setting 2 – Varying the Number of Subtasks: First, simulations considered agents with capabilities to play any role. Figure 3a shows that although the performance decreased somewhat with more subtasks, it is still close to the optimal solution. Even though the distance between the coefficients of variation increases with more subtasks, the distance still represent small differences (Fig. 3b). The average number of bids that each agent provides increase with more subtasks while the average number of bids by subtask decreases (Fig. 3c). Then for the next simulations we randomly assign from 1 to 4 capabilities to each robot. Figure 4 shows the results with similar performance obtained in the first part of the simulations, even when the capabilities of the robots limit the tasks that they can perform. The average execution time was 5 s for each simulation with 15 subtasks, and 26 s with 60 subtasks.

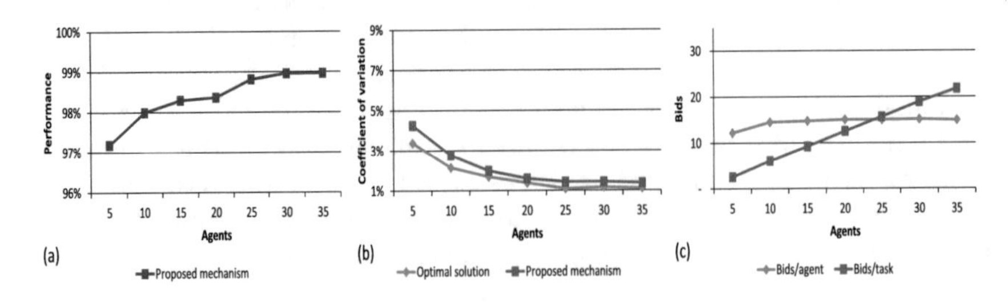

Fig. 1. Performance results by varying the number of agents.

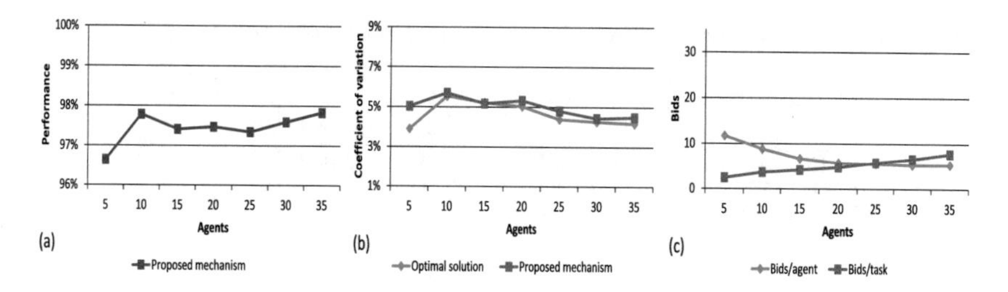

Fig. 2. Performance results by varying the number of agents and agents capabilities.

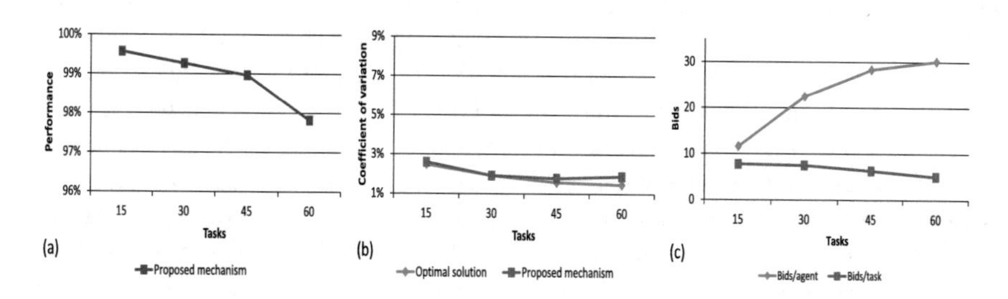

Fig. 3. Performance results by varying the number of subtasks.

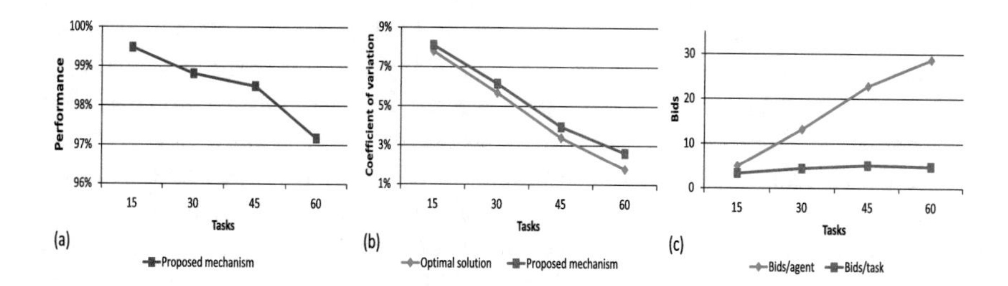

Fig. 4. Performance results by varying the number of subtasks and agents capabilities.

Setting 3 – Coping with Partially Allocated Tasks: These simulations considered that the number of subtasks was greater than the total capacity of the agents. Thus, at the end of allocation we may have tasks partially allocated and others completely unallocated, which the agents will try to reallocated. Table 2 shows reasonable performance results on reallocating partially allocated tasks, where PA is the number of partially allocated tasks, NA is the number of completely unallocated tasks and EA represents the entirely allocated tasks (i.e., all subtasks have been allocated). The execution time varied from 3 to 10 s.

Table 2. Simulations when number of subtasks is greater than total capacity of robots

Tasks	Subtasks		PA	NA	EA		PA	NA	EA
9	21	First allocation	3	1	5	Second allocation	0	3	6
12	28		5	4	3		0	6	6
15	35		6	5	4		0	10	5

5 Related Work

There are several works on tasks allocation, some them aims at allocating an initial set of tasks to a set of robots [2,5,9], while others focus on allocation of tasks that arise during the execution of other tasks [4,10]. [9], for instance, presents a distributed solution for task allocation to heterogeneous robots in which robots' capabilities are considered. Unlike our proposal, the solution requires that only one task is allocated to each robot and focus only on atomic tasks. Further, in [2] is introduced a centralised approach for task allocation that considers heterogeneous robots with different capabilities. However, each robot can be allocated only two subtasks at the same time (current and next subtask to be carried out). In [4] a framework for allocating new tasks discovered by robots during their missions is put forward. It proposes the use of heterogeneous robots, in which the robot with the best computational resources is responsible for the allocation process. Thus, it could be said of that there is still a single point of failure within each team, so it is not exactly a decentralised solution like ours. The work presented in [5] is the basis for the mechanism proposed here. It presents a distributed algorithm focusing on the allocation of groups of tasks. There are constraints in the total number of tasks that a robot can carry out in the mission. It is assumed that any robot can be allocated to any task. Unlike us, the work does not consider the allocation of different types of tasks at the same time, aspects related to capabilities of robots, and the use of roles associated with tasks is not considered either.

6 Conclusion

We proposed a decentralised mechanism for the dynamic allocation of different types of tasks to heterogeneous robots, considering that they can play different roles and carry out tasks according to their own capabilities. The mechanism allows to represent type of tasks such as DS, CN and CM and also to express other constraints through the definition of minimum and maximum values.

The results demonstrate that the proposed mechanism seems to scale well, as well as provides near-optimal allocations. However, the number of bid messages exchanged between the robots could impact the solution in a real-world environment. Future work aims to consider aspects such as task prioritising and other task constraints. We also intend to evaluate the approach with real robots or at least more realistic settings (e.g., including details of available communication links for robots) on ROS-based simulations that are currently being developed.

Acknowledgements. We acknowledge the support given by CNPQ and CAPES/Pro-Alertas (88887.115590/2015-01). Tulio Basegio thanks the support given by Federal Institute of Rio Grande do Sul (IFRS).

References

1. Boissier, O., Bordini, R.H., Hübner, J.F., Ricci, A., Santi, A.: Multi-agent oriented programming with jacamo. Sci. Comput. Program. **78**(6), 747–761 (2013)
2. Das, G.P., McGinnity, T.M., Coleman, S.A.: Simultaneous allocations of multiple tightly-coupled multi-robot tasks to coalitions of heterogeneous robots. In: 2014 IEEE International Conference on Robotics and Biomimetics (ROBIO), pp. 1198–1204 (2014)
3. Gerkey, B., Mataric, M.: A formal analysis and taxonomy of task allocation in multi-robot systems. Int. J. Robot. Res. **23**(9), 939–954 (2004)
4. Gunn, T., Anderson, J.: Effective task allocation for evolving multi-robot teams in dangerous environments. In: 2013 IEEE/WIC/ACM International Joint Conference on Web Intelligence (WI) and Intelligent Agent Technologies (IAT), vol. 2, pp. 231–238 (2013)
5. Luo, L., Chakraborty, N., Sycara, K.: Provably-good distributed algorithm for constrained multi-robot task assignment for grouped tasks. IEEE Trans. Robot. **31**(1), 19–30 (2015)
6. Murphy, R.R., Tadokoro, S., Nardi, D., Jacoff, A., Fiorini, P., Choset, H., Erkmen, A.M.: Search and Rescue Robotics, pp. 1151–1173. Springer, Heidelberg (2008)
7. Ramchurn, S.D., et al.: Hacer: a disaster response system based on human-agent collectives. In: Proceedings of the 2015 International Conference on Autonomous Agents and Multiagent Systems, AAMAS 2015, pp. 533–541. International Foundation for Autonomous Agents and Multiagent Systems, Richland, SC (2015)
8. Scerri, P., et al.: Flood disaster mitigation: a real-world challenge problem for multi-agent unmanned surface vehicles. In: Dechesne, F., Hattori, H., Mors, A., Such, J.M., Weyns, D., Dignum, F. (eds.) AAMAS 2011. LNCS (LNAI), vol. 7068, pp. 252–269. Springer, Heidelberg (2012). doi:10.1007/978-3-642-27216-5_16
9. Settimi, A., Pallottino, L.: A subgradient based algorithm for distributed task assignment for heterogeneous mobile robots. In: 52nd IEEE Conference on Decision and Control, pp. 3665–3670 (2013)
10. Urakawa, K., Sugawara, T.: Task allocation method combining reorganization of agent networks and resource estimation in unknown environments. In: 2013 Third International Conference on Innovative Computing Technology (INTECH), pp. 383–388 (2013)
11. Yan, Z., Jouandeau, N., Cherif, A.A.: A survey and analysis of multi-robot coordination. Int. J. Adv. Robot. Syst. **10**, 1–10 (2013)
12. Zlot, R.M.: An auction-based approach to complex task allocation for multirobot teams. Ph.D. thesis, Robotics Institute, Carnegie Mellon University (2006)
13. Makhorin, A.: GNU Linear Programming Kit reference manual. http://www.gnu.org/software/glpk/glpk.html

SAT-Versus SMT-Based BMC for TWIS and the Existential Fragment of WCTL with Knowledge

Agnieszka M. Zbrzezny(✉)

IMCS, Jan Długosz University, Al. Armii Krajowej 13/15,
42-200 Częstochowa, Poland
agnieszka.zbrzezny@ajd.czest.pl

Abstract. In this paper, we present the SAT-based bounded model checking method for Timed Weighted Interpreted Systems and for Weighted Existential Computation Tree Logic with epistemic operators. SAT-based bounded model checking consists in translating the existential model checking problem for a modal logic and for a model to the boolean satisfiability problem. We provide an implementation based on Cryptominisat and YicesSAT SAT-solvers and we present a comparison of the SAT-based BMC method and SMT-based BMC methods on common instances that can be scaled up to for performance evaluation.

1 Introduction

Multi-agent systems (MASs) are composed of many intelligent agents that interact with each other. The agents can share a common goal or they can pursue their own interests. Also, the agents may have a deadline or other timing constraints to achieve intended targets. As it was shown in [3], knowledge is a useful concept for analysing the information state and the behaviour of agents in multi-agent systems.

Model checking [1] is an automatic verification technique for concurrent systems. To be able to check automatically whether the system satisfies a given property, one must first create a model of the system, and then describe in a formal language both the created model and the property. One of the main technique here is the *symbolic model checking* [1]. Unfortunately, because of the agents' intricate nature, the practical applicability of model checking is firmly limited by the "state-space explosion problem". To reduce this issue, various techniques, including the SAT- and BDD-based bounded model checking (BMC) [5,8], have been advanced. These have been effective in permitting users to handle bigger MASs, however it is still hard to check MASs with numerous agents.

Partly supported by National Science Centre under the grant No. 2014/15/N/ST6/05079.

G. Jezic et al. (eds.), *Agent and Multi-Agent Systems: Technology and Applications*, Smart Innovation, Systems and Technologies 74,
DOI 10.1007/978-3-319-59394-4_11

Timed weighted interpreted systems (TWIS) were proposed in [13] to extend interpreted systems in order to make possible reasoning about real-time aspects of MASs and to make the reasoning possible about not only temporal and epistemic properties, but also agents's quantitative properties. In this paper, we consider the existential fragment of a weighted epistemic computation tree logic (WECTLK) interpreted over TWISs. We propose SAT-based bounded model checking technique for this kind of systems and specifications described in WECTLK.

The original contributions of the paper are as follows. First, we propose a SAT-based BMC technique for TWIS and for WECTLK- the SAT-based method is much more difficult to define and implement than SMT-based method. We also improved the translation of the abstract model presented in [13]. Second, we report on the implementation of the proposed BMC method as a new module of a verification system, and evaluate it experimentally by means of a modified *generic pipeline paradigm* [7]. We compare our new method with the method presented in [13] because it is the only method which supports the WECTLK language and the timed weighted interpreted systems, and we improved the old implementation of the SMT-based BMC method.

2 Preliminaries

Let $\mathbb{N}$ be a set of natural numbers, and $\mathbb{N}_+ = \mathbb{N} \setminus \{0\}$. We assume a finite set $\mathcal{X}$ of variables, called *clocks*. Each clock is a variable ranging over a set of non-negative natural numbers. For $x \in \mathcal{X}$, $\bowtie \in \{<, \leq, =, >, \geq\}$, $c \in \mathbb{N}$ we define a set of clock constraints over $\mathcal{X}$, denoted by $\mathcal{C}(\mathcal{X})$, The constraints are conjunctions of comparisons of a clock with a time constant c from the set of natural numbers $\mathbb{N}$, generated by the following grammar: $\mathfrak{cc} := \mathbf{true} \mid x \bowtie c \mid \mathfrak{cc} \wedge \mathfrak{cc}$. A clock valuation v of $\mathcal{X}$ is a total function from $\mathcal{X}$ into the set of natural numbers. The set of all the clock valuations is denoted by $\mathbb{N}^{\mathcal{X}}$. For $\mathcal{X}' \subseteq \mathcal{X}$, the valuation which assigns the value 0 to all clocks is defined as: $\forall_{x \in \mathcal{X}'} v'(x) = 0$ and $\forall_{x \in \mathcal{X} \setminus \mathcal{X}'} v'(x) = v(x)$. For $v \in \mathbb{N}^{\mathcal{X}}$, $succ(v)$ is the clock valuation of $\mathcal{X}$ that assigns the value $v(x) + 1$ to each clock x. A clock valuation v satisfies a clock constraint $\mathfrak{cc}$, written as $v \models \mathfrak{cc}$, iff $\mathfrak{cc}$ evaluates to true using the clock values given by v.

TWISs. Let $\mathcal{A} = \{1, \ldots, n\}$ denote a non-empty and finite set of agents, and $\mathcal{E}$ be a special agent that is used to model the environment in which the agents operate, and $\mathcal{AP} = \bigcup_{\mathbf{i} \in \mathcal{A} \cup \{\mathcal{E}\}} \mathcal{AP}_{\mathbf{i}}$ be a set of atomic formulae, such that $\mathcal{AP}_{\mathbf{i}_1} \bigcap \mathcal{AP}_{\mathbf{i}_2} = \emptyset$ for all $\mathbf{i}_1, \mathbf{i}_2 \in \mathcal{A} \cup \{\mathcal{E}\}$.

A *timed interpreted system* is a tuple $TWIS = (\{L_{\mathbf{i}}, Act_{\mathbf{i}}, \mathcal{X}_{\mathbf{i}}, P_{\mathbf{i}}, \mathcal{V}_{\mathbf{i}}, \mathcal{I}_{\mathbf{i}}, \iota_{\mathbf{i}}, d_{\mathbf{i}}\}_{\mathbf{i} \in \mathcal{A} \cup \{\mathcal{E}\}}, \{t_{\mathbf{i}}\}_{\mathbf{i} \in \mathcal{A}}, \{t_{\mathcal{E}}\})$, where: $L_{\mathbf{i}}$ is a non-empty set of *locations* of the agent $\mathbf{i}$, $\iota_{\mathbf{i}} \subseteq L_{\mathbf{i}}$ is a non-empty set of initial locations, $Act_{\mathbf{i}}$ is a non-empty set of *possible actions* of the agent $\mathbf{i}$, $Act = Act_1 \times \ldots \times Act_n \times Act_{\mathcal{E}}$ is the set of *joint actions*, $\mathcal{X}_{\mathbf{i}}$ is a non-empty set of *clocks*, $P_{\mathbf{i}} : L_{\mathbf{i}} \rightarrow 2^{Act_{\mathbf{i}}}$ is a *protocol function*, $t_{\mathbf{i}} : L_{\mathbf{i}} \times L_{\mathcal{E}} \times \mathcal{C}(\mathcal{X}_{\mathbf{i}}) \times 2^{\mathcal{X}_{\mathbf{i}}} \times Act \rightarrow L_{\mathbf{i}}$ is a (partial) *evolution function* for agents, $t_{\mathcal{E}} : L_{\mathcal{E}} \times \mathcal{C}(\mathcal{X}_{\mathcal{E}}) \times 2^{\mathcal{X}_{\mathcal{E}}} \times Act \rightarrow L_{\mathcal{E}}$ is a (partial) *evolution function* for environment, $\mathcal{V}_{\mathbf{i}} : L_{\mathbf{i}} \rightarrow 2^{\mathcal{AP}_{\mathbf{i}}}$ is a *valuation function* assigning to each location a set of atomic formulae that are assumed to be true at that location, $\mathcal{I}_{\mathbf{i}}$: $L_{\mathbf{i}} \rightarrow \mathcal{C}(\mathcal{X}_{\mathbf{i}})$ is an *invariant function*, that specifies the amount of time the agent $\mathbf{i}$ may spend in a given location, $d_{\mathbf{i}} : Act_{\mathbf{i}} \rightarrow \mathbb{N}$ is a *weight function.* It is assumed that locations, actions and clocks for the environment are "public", which means that all the agents know a current location, an action, and a clock valuation of the environment. We also assume that if $\epsilon_{\mathbf{i}} \in P_{\mathbf{i}}(l_{\mathbf{i}})$, then $t_{\mathbf{i}}(l_{\mathbf{i}}, l_{\mathcal{E}}, \mathfrak{cc}_{\mathbf{i}}, \mathcal{X}, (a_1, \ldots, a_n, a_{\mathcal{E}})) = l_{\mathbf{i}}$ for $a_{\mathbf{i}} = \epsilon_{\mathbf{i}}$, any $\mathfrak{cc}_{\mathbf{i}} \in \mathcal{C}(\mathcal{X}_{\mathbf{i}})$, and any $\mathcal{X} \subseteq \mathcal{X}_{\mathbf{i}}$. Each element t of $t_{\mathbf{i}}$ is denoted by $< l_{\mathbf{i}}, l_{\mathcal{E}}, \mathfrak{cc}_{\mathbf{i}}, \mathcal{X}', a, l'_{\mathbf{i}} >$, where $l_{\mathbf{i}}$ is the source location, $l'_{\mathbf{i}}$ is the target location, a is an action, $\mathfrak{cc}$ is the enabling condition for $t_{\mathbf{i}}$, and $\mathcal{X}' \subseteq \mathcal{X}_{\mathbf{i}}$ is the set of clocks to be reset after performing t. An invariant condition allows the TWIS to stay at the location l as long as the constraint $\mathcal{I}_{\mathbf{i}}(l_{\mathbf{i}})$ is satisfied. The guard $\mathfrak{cc}$ has to be satisfied to enable the transition.

Timed Weighted Model. For a given TWIS let the symbol $S = \prod_{\mathbf{i} \in \mathcal{A} \cup \{\mathcal{E}\}}(L_{\mathbf{i}} \times \mathbb{N}^{\mathcal{X}_{\mathbf{i}}})$ denote the non-empty set of all *global states*. Moreover, for a given global state $s = ((l_1, v_1), \ldots, (l_n, v_n), (l_{\mathcal{E}}, v_{\mathcal{E}})) \in S$, let the symbols $l_{\mathbf{i}}(s) = l_{\mathbf{i}}$ and $v_{\mathbf{i}}(s) = v_{\mathbf{i}}$ denote, respectively, the local component and the clock valuation of agent $\mathbf{i} \in \mathcal{A} \cup \{\mathcal{E}\}$ in s. Now, for a given TWIS we define a *timed model* (or a *model*) as a tuple $\mathcal{M} = (S, Act, \iota, T, \mathcal{V})$, where: $Act = Act_1 \times \ldots \times Act_n \times Act_{\mathcal{E}}$ is the set of all the joint actions, $S = \prod_{\mathbf{i} \in \mathcal{A} \cup \{\mathcal{E}\}}(L_{\mathbf{i}} \times \mathbb{N}^{\mathcal{X}_{\mathbf{i}}})$ is the set of all the *global states*, $\iota = \prod_{\mathbf{i} \in \mathcal{A} \cup \{\mathcal{E}\}}(\iota_i \times \{0\}^{\mathcal{X}_i})$ is the set of all the *initial* global states, $\mathcal{V} : S \rightarrow 2^{\mathcal{AP}}$ is the valuation function defined as $\mathcal{V}(s) = \bigcup_{\mathbf{i} \in \mathcal{A} \cup \{\mathcal{E}\}} \mathcal{V}_{\mathbf{i}}(l_{\mathbf{i}}(s))$, $T \subseteq S \times (Act \cup \mathbb{N}) \times S$ is a transition relation defined by action and time transitions. For $\widetilde{a} \in Act$:

1. action transition: $(s, \widetilde{a}, s') \in T$ (or $s \xrightarrow{\widetilde{a}} s'$) iff for all $\mathbf{i} \in \mathcal{A} \cup \{\mathcal{E}\}$, there exists a local transition $t_{\mathbf{i}}(l_{\mathbf{i}}(s), \mathfrak{cc}_{\mathbf{i}}, \mathcal{X}', \widetilde{a}) = l_{\mathbf{i}}(s')$ such that $v_{\mathbf{i}}(s) \models \mathfrak{cc}_{\mathbf{i}} \wedge \mathcal{I}(l_{\mathbf{i}}(s))$ and $v'_{\mathbf{i}}(s') = v_{\mathbf{i}}(s)[\mathcal{X}' := 0]$ and $v'_{\mathbf{i}}(s') \models \mathcal{I}(l_{\mathbf{i}}(s'))$ ($v_{\mathbf{i}}(s)[\mathcal{X}' := 0]$ denotes the clock valuation which assigns 0 to each clock in $\mathcal{X}'$ and agrees with $v_{\mathbf{i}}(s)$ over the rest of the clocks.
2. time transition $(s, \delta, s') \in T$ iff for all $\mathbf{i} \in \mathcal{A} \cup \{\mathcal{E}\}$, $l_{\mathbf{i}}(s) = l_{\mathbf{i}}(s')$ and $v'_{\mathbf{i}}(s') = v_{\mathbf{i}}(s) + \delta$ and $v'_{\mathbf{i}}(s') \models \mathcal{I}(l_{\mathbf{i}}(s'))$.

The "joint" weight function $d : Act \rightarrow \mathbb{N}$ is defined as follows: $d((a_1, \ldots, a_n, a_{\mathcal{E}})) = d_1(a_1) + \ldots + d_n(a_n) + d_{\mathcal{E}}(a_{\mathcal{E}})$.

Given a timed weighted model, one can define for any agent $\mathbf{i}$ the *indistinguishability* relation $\sim_{\mathbf{i}} \subseteq S \times S$ as follows: $s \sim_{\mathbf{i}} s'$ iff $l_{\mathbf{i}}(s') = l_{\mathbf{i}}(s)$ and $v_{\mathbf{i}}(s') = v_{\mathbf{i}}(s)$. We assume the following definitions of epistemic relations: $\sim^E_\Gamma \stackrel{def}{=} \bigcup_{\mathbf{i} \in \Gamma} \sim_{\mathbf{i}}$, $\sim^C_\Gamma \stackrel{def}{=} (\sim^E_\Gamma)^+$ (the transitive closure of $\sim^E_\Gamma$), $\sim^D_\Gamma \stackrel{def}{=} \bigcap_{\mathbf{i} \in \Gamma} \sim_{\mathbf{i}}$, where $\Gamma \subseteq \mathcal{A}$.

A run in $\mathcal{M}$ is an infinite sequence $\rho = s_0 \xrightarrow{\delta_0, \widetilde{a}_0} s_1 \xrightarrow{\delta_1, \widetilde{a}_1} s_2 \xrightarrow{\delta_2, \widetilde{a}_2} \ldots$ of global states such that the following conditions hold for all $i \in \mathbb{N}$: $s_i \in S, \widetilde{a}_i \in Act, \delta_i \in \mathbb{N}_+$, and there exists $s'_i \in S$ such that $(s_i, \delta, s'_i) \in T$ and $(s_i, \widetilde{a}, s_{i+1}) \in T$. Notice that the definition of a run does not permit two consecutive joint actions to be performed one after the other, i.e., between each two joint actions some time must pass; such a run is called *strongly monotonic*.

Abstract model. Let $\mathbb{D}_{\mathbf{i}} = \{0, \ldots, c_{\mathbf{i}} + 1\}$ with $c_{\mathbf{i}}$ be the largest constant appearing in any enabling condition or state invariants of agent $\mathbf{i}$ and $\mathbb{D} = \bigcup_{\mathbf{i} \in \mathcal{A} \cup \mathcal{E}} \mathbb{D}_{\mathbf{i}}^{|\mathcal{X}_{\mathbf{i}}|}$. A tuple $\widehat{\mathcal{M}} = (\widehat{S}, Act, \widehat{\iota}, \widehat{T}, \widehat{\mathcal{V}}, d)$, is an *abstract model*, where $\widehat{\iota} = \prod_{\mathbf{i} \in \mathcal{A} \cup \mathcal{E}} \iota_{\mathbf{i}} \times \{0\}^{|\mathcal{X}_{\mathbf{i}}|}$ is the set of all the initial global states, $\widehat{S} = \prod_{\mathbf{i} \in \mathcal{A} \cup \mathcal{E}} L_{\mathbf{i}} \times \mathbb{D}_{\mathbf{i}}^{|\mathcal{X}_{\mathbf{i}}|}$ is the set of all the abstract global states. $\widehat{\mathcal{V}} : \widehat{S} \to 2^{\mathcal{AP}}$ is the valuation function such that: $p \in \widehat{\mathcal{V}}(\widehat{s})$ iff $p \in \bigcup_{\mathbf{i} \in \mathcal{A} \cup \mathcal{E}} \widehat{\mathcal{V}}_{\mathbf{i}}(l_{\mathbf{i}}(\widehat{s}))$ for all $p \in \mathcal{AP}$; and $\widehat{T} \subseteq \widehat{S} \times (Act \cup \tau) \times \widehat{S}$. Let $\widetilde{a} \in Act$. Then,

1. Action transition: $(\widehat{s}, \widetilde{a}, \widehat{s}') \in \widehat{T}$ iff $\forall_{\mathbf{i} \in \mathcal{A}} \exists_{\phi_{\mathbf{i}} \in \mathcal{C}(\mathcal{X}_{\mathbf{i}})} \exists_{\mathcal{X}_{\mathbf{i}}' \subseteq \mathcal{X}_{\mathbf{i}}} (t_{\mathbf{i}}(l_{\mathbf{i}}(\widehat{s}), \phi_{\mathbf{i}}, \mathcal{X}_{\mathbf{i}}', \widetilde{a}) = l_{\mathbf{i}}(\widehat{s}')$ and $v_{\mathbf{i}} \models \phi_{\mathbf{i}} \wedge \mathcal{I}(l_{\mathbf{i}}(\widehat{s}))$ and $v_{\mathbf{i}}'(\widehat{s}') = v_{\mathbf{i}}(\widehat{s})[\mathcal{X}_{\mathbf{i}}' := 0]$ and $v_{\mathbf{i}}'(\widehat{s}') \models \mathcal{I}(l_{\mathbf{i}}(\widehat{s}')))$
2. Time transition: $(\widehat{s}, \tau, \widehat{s}') \in \widehat{T}$ iff $\forall_{\mathbf{i} \in \mathcal{A} \cup \mathcal{E}} (l_{\mathbf{i}}(\widehat{s}) = l_{\mathbf{i}}(\widehat{s}'))$ and $v_{\mathbf{i}}(\widehat{s}) \models \mathcal{I}(l_{\mathbf{i}}(\widehat{s}))$ and $succ(v_{\mathbf{i}}(\widehat{s})) \models \mathcal{I}(l_{\mathbf{i}}(\widehat{s})))$ and $\forall_{\mathbf{i} \in \mathcal{A}} (v_{\mathbf{i}}'(\widehat{s}') = succ(v_{\mathbf{i}}(\widehat{s}')))$ and $(v_{\mathcal{E}}'(\widehat{s}') = succ(v_{\mathcal{E}}(\widehat{s})))$.

Given the abstract model one can define for any agent $\mathbf{i}$ the indistinguishability relation $\sim_{\mathbf{i}} \subseteq \widehat{S} \times \widehat{S}$ as follows: $\widehat{s} \sim_{\mathbf{i}} \widehat{s}'$ iff $l_{\mathbf{i}}(\widehat{s}') = l_{\mathbf{i}}(\widehat{s})$ and $v_{\mathbf{i}}(\widehat{s}') = v_{\mathbf{i}}(\widehat{s})$. A run ρ in the abstract model is a sequence $\widehat{s}_0 \xrightarrow{b_1} \widehat{s}_1 \xrightarrow{b_2} \widehat{s}_2 \xrightarrow{b_3} \ldots$ of transitions such that for each $i \leq 1$, $b_i \in Act \cup \{\tau\}$ and $b_1 = \tau$ and for each two consecutive transitions at least one of them is a time transition. Next, $\rho[j..m]$ denotes the finite sequence $\widehat{s}_j \xrightarrow{\delta_{j+1}, \widetilde{a}_{j+1}} \widehat{s}_{j+1} \xrightarrow{\delta_{j+2}, \widetilde{a}_{j+2}} \ldots \widehat{s}_m$ with $m - j$ transitions and $m - j + 1$ states, and $\rho(j)$ denotes j-th state at the run ρ. $D\rho[j..m]$ denotes the (cumulative) weight of $\rho[j..m]$ that is defined as $d(\widetilde{a}_{j+1}) + \ldots + d(\widetilde{a}_m)$ (hence 0 when $j = m$). The set of all the runs starting at $\widehat{s} \in \widehat{S}$ is denoted by $\mathrm{P}(\widehat{s})$, and the set of all the runs starting at an initial state is denoted by $\mathrm{P} = \bigcup_{\widehat{s}^0 \in \widehat{\iota}} \mathrm{P}(\widehat{s}^0)$.

WECTLK. The WECTLK has been defined in [9] as the existential fragment of the weighted CTLK with integer cost constraints on all temporal modalities. In the syntax of WECTLK we assume the following: $p \in \mathcal{AP}$ is an atomic proposition, $\mathbf{i} \in \mathcal{A}$, $\Gamma \subseteq \mathcal{A}$, I is an interval in $\mathbb{N}$ of the form: $[a, \infty)$ and $[a, b)$, for $a, b \in \mathbb{N}$ and $a \neq b$. Moreover, hereafter, $\mathbf{right}(I)$ denotes the right end of the interval I. The WECTLK formulae are defined by the following grammar:

$$\varphi ::= \mathbf{true} \,|\, \mathbf{false} \,|\, p \,|\, \neg p \,|\, \varphi \vee \varphi \,|\, \varphi \wedge \varphi \,|\, \mathbf{EX}_I \varphi \,|\, \mathbf{E}(\varphi \mathbf{U}_I \varphi) \,|\, \mathbf{EG}_I \varphi \,|\, \overline{\mathrm{K}}_{\mathbf{i}} \varphi.$$

A WECTLK formula φ is *true* in the abstract model $\widehat{\mathcal{M}}$ (in symbols $\widehat{\mathcal{M}} \models \varphi$) iff $\widehat{\mathcal{M}}, \widehat{s}^0 \models \varphi$ for some $\widehat{s}^0 \in \widehat{\iota}$ (i.e., φ is true at some initial state of the abstract model $\widehat{\mathcal{M}}$). For every $\widehat{s} \in \widehat{S}$ the relation $\models$ is defined inductively as follows:

- $\widehat{\mathcal{M}}, \widehat{s} \models \mathbf{true}$, $\widehat{\mathcal{M}}, \widehat{s} \not\models \mathbf{false}$, $\widehat{\mathcal{M}}, \widehat{s} \models p$ iff $p \in \widehat{\mathcal{V}}(\widehat{s})$, $\widehat{\mathcal{M}}, \widehat{s} \models \neg p$ iff $p \notin \widehat{\mathcal{V}}(\widehat{s})$,
- $\widehat{\mathcal{M}}, \widehat{s} \models \alpha \wedge \beta$ iff $\widehat{\mathcal{M}}, \widehat{s} \models \alpha$ and $\widehat{\mathcal{M}}, \widehat{s} \models \beta$,
- $\widehat{\mathcal{M}}, \widehat{s} \models \alpha \vee \beta$ iff $\widehat{\mathcal{M}}, \widehat{s} \models \alpha$ or $\widehat{\mathcal{M}}, \widehat{s} \models \beta$
- $\widehat{\mathcal{M}}, \widehat{s} \models \mathbf{EX}_I \alpha$ iff $(\exists \rho \in \mathrm{P}(\widehat{s}))(D\rho[0..1] \in I$ and $\widehat{\mathcal{M}}, \rho(1) \models \alpha)$,
- $\widehat{\mathcal{M}}, \widehat{s} \models \mathbf{EG}_I \alpha$ iff $(\exists \rho \in \mathrm{P}(\widehat{s}))(\forall i \geq 0)(D\rho[0..i] \in I$ implies $\widehat{\mathcal{M}}, \rho(i) \models \beta)$,
- $\widehat{\mathcal{M}}, \widehat{s} \models \mathbf{E}(\alpha \mathbf{U}_I \beta)$ iff $(\exists \rho \in \mathrm{P}(\widehat{s}))(\exists i \geq 0)(D\rho[0..i] \in I$ and $\widehat{\mathcal{M}}, \rho(i) \models \beta$ and $(\forall j < i) \widehat{\mathcal{M}}, \rho(j) \models \alpha)$,
- $\widehat{\mathcal{M}}, \widehat{s} \models \overline{\mathrm{K}}_{\mathbf{i}} \alpha$ iff $(\exists \rho \in \mathrm{P})$ $(\exists i \geq 0)(\widehat{s} \sim_{\mathbf{i}} \rho(i)$ and $\widehat{\mathcal{M}}, \rho(i) \models \alpha)$.

The *model checking problem* asks whether $\widehat{\mathcal{M}} \models \varphi$. Note that the formula "weighted eventually" is defined as standard: $\mathbf{EF}_I\varphi \stackrel{def}{=} \mathbf{E}(\mathbf{true}\mathbf{U}_I\varphi)$ (meaning that it is possible to reach a state satisfying φ via a finite run whose cumulative weight is in I).

3 SAT-based Bounded Model Checking

In this section, we present an outline of the bounded semantics for WECTLK and define an SAT-based BMC method for WECTLK, which is based on the BMC encoding presented in [9]. As usual, we start by defining k-runs and $(k,l)-loops$. Next, we define a bounded semantics, which is used for the translation to SAT.

Bounded semantics. Let $\widehat{\mathcal{M}}$ be the abstract model for TWIS, and $k \in \mathbb{N}$ a bound. A *k-run* ρ_k is a finite sequence $\widehat{s}_0 \xrightarrow{b_1} \widehat{s}_1 \xrightarrow{b_2} \dots \xrightarrow{b_k} \widehat{s}_k$ of transitions such that for each $1 \leq i \leq k$, $b_i \in Act \cup \{\tau\}$ and $b_1 = \tau$ and for each two consecutive transitions at least one is a time transition. A k-run ρ_k is a *loop* if $l < k$ and $\rho(k) = \rho(l)$. Note that if a k-run ρ_k is a loop, then it represents the infinite run of the form uv^ω, where $u = (s_0 \xrightarrow{b_1} s_1 \xrightarrow{b_2} \dots \xrightarrow{b_l} s_l)$ and $v = (s_{l+1} \xrightarrow{b_{l+2}} \dots \xrightarrow{b_k} s_k)$. $\mathrm{P}_k(\widehat{s})$ denotes the set of all the k-runs of $\widehat{\mathcal{M}}$ that start at $\widehat{s}$, and $\mathrm{P}_k = \bigcup_{\widehat{s}^0 \in \widehat{\iota}} \mathrm{P}_k(\widehat{s}^0)$.

The bounded satisfiability relation $\models_k$ which indicates k-truth of a WECTLK formula in the abstract model $\widehat{\mathcal{M}}$ at some state $\widehat{s}$ of $\widehat{\mathcal{M}}$ is also defined in [9]. A WECTLK formula φ is *k-true* in the abstract model $\widehat{\mathcal{M}}$ (in symbols $\widehat{\mathcal{M}} \models_k \varphi$) iff φ is k-true at some initial state of the abstract model $\widehat{\mathcal{M}}$.

The *bounded model checking problem* asks whether there exists $k \in \mathbb{N}$ such that $\widehat{\mathcal{M}} \models_k \varphi$. The following theorem states that for a given abstract model and a WECTLK formula there exists a bound k such that the model checking problem ($\widehat{\mathcal{M}} \models \varphi$) can be reduced to the bounded model checking problem ($\widehat{\mathcal{M}} \models_k \varphi$).

Theorem 1. *Let $\widehat{\mathcal{M}}$ be the abstract model and φ a WECTLK formula. Then, the following equivalence holds: $\widehat{\mathcal{M}} \models \varphi$ iff there exists $k \geq 0$ such that $\widehat{\mathcal{M}} \models_k \varphi$.*

Proof. The theorem can be proven by induction on the length of the formula φ (for details one can see [11]).

Translation to SAT. Let $\widehat{\mathcal{M}}$ be the abstract model for TWIS, φ a WECTLK formula, and $k \geq 0$ a bound. The presented SAT encoding of the BMC problem for WECTLK and for TWIS is based on the SAT encoding of the same problem [10,12], and it relies on defining the propositional formula: $[\widehat{\mathcal{M}}, \varphi]_k := [\widehat{\mathcal{M}}^{\varphi,\widehat{\iota}}]_k \wedge [\varphi]_{\widehat{\mathcal{M}},k}$ that is satisfiable if and only if $\widehat{\mathcal{M}} \models_k \varphi$ holds.

Let $\mathbf{i} \in \mathcal{A} \cup \{\mathcal{E}\}$. The definition of the formula $[\widehat{\mathcal{M}}, \varphi]_k$ assumes that each global state $s \in \widehat{S}$ is represented by a valuation of a *symbolic state*

$\overline{\mathbf{w}} = ((w_1, v_1), \ldots, (w_n, v_n), (w_\mathcal{E}, v_\mathcal{E}))$ that consists of *symbolic local states* and each symbolic local state $w_\mathbf{i}$ is a pair $(w_\mathbf{i}, v_\mathbf{i})$ of individual variables ranging over the natural numbers, in which the first element represents a local state of the agent **i**, and the second represents a clock valuation; each joint action $\widetilde{a} \in Act$ is represented by a valuation of a *symbolic action* $\overline{a} = (a_1, \ldots, a_n, a_\mathcal{E})$ that consists of *symbolic local actions* and each symbolic local action $a_\mathbf{i}$ is an individual variable ranging over the natural numbers; each sequence of weights associated with the joint action is represented by a valuation of a *symbolic weights* $\overline{d} = (d_1, \ldots, d_{n+1})$ that consists of *symbolic local weights* and each symbolic local weight $d_\mathbf{i}$ is an individual variable ranging over the natural numbers.

The formula $[\widehat{\mathcal{M}}^{\varphi,\widehat{\iota}}]_k$ encodes a rooted tree of k-runs of the abstract model $\widehat{\mathcal{M}}$. The number of branches of the tree depends on the value of $f_k : \text{WECTLK} \rightarrow \mathbb{N}$ which is an auxiliary function defined in [9]. The formula $[\widehat{\mathcal{M}}^{\varphi,\widehat{\iota}}]_k$ is defined over $(k+1) \cdot f_k(\varphi)$ different symbolic states, $k \cdot f_k(\varphi)$ different symbolic actions, and $k \cdot f_k(\varphi)$ different symbolic weights. Moreover, it uses the following auxiliary propositional formulae:

- $I_s(\overline{\mathbf{w}})$ - it encodes the state s of the abstract model $\widehat{\mathcal{M}}$;
- $H_\mathbf{i}(w_\mathbf{i}, w'_\mathbf{i})$ - it encodes equality of two local states, such that $w_\mathbf{i} = w'_\mathbf{i}$ for $\mathbf{i} \in \mathcal{A} \cup \mathcal{E}$;
- $\mathcal{T}_\mathbf{i}(w_\mathbf{i}, ((\overline{a}, \overline{d}), \overline{\delta}), w'_\mathbf{i})$ - it encodes the local evolution function of agent **i**;
- $\mathcal{A}(\overline{a})$ - it encodes that each symbolic local action $a_\mathbf{i}$ of $\overline{a}$ has to be executed by each agent in which it appears;
- $\mathcal{T}(\overline{\mathbf{w}}, ((\overline{a}, \overline{d}), \overline{\delta}), \overline{\mathbf{w}}') := \mathcal{A}(\overline{a}) \wedge \bigwedge_{\mathbf{i} \in \mathcal{A} \cup \{\mathcal{E}\}} \mathcal{T}_\mathbf{i}(w_\mathbf{i}, ((\overline{a}, \overline{d}), \overline{\delta}), w'_\mathbf{i})$;
- Let $\boldsymbol{\rho}_j$ denote the j-th *symbolic k-run*, i.e. the sequence of symbolic transitions: $\overline{\mathbf{w}}_{0,j} \overset{(\overline{a}_{1,j}, \overline{d}_{1,j}), \delta_{1,j}}{\longrightarrow} \overline{\mathbf{w}}_{1,j} \overset{(\overline{a}_{2,j}, \overline{d}_{2,j}), \delta_{2,j}}{\longrightarrow} \ldots \overset{(\overline{a}_{k,j}, \overline{d}_{k,j}), \delta_{k,j}}{\longrightarrow} \overline{\mathbf{w}}_{k,j}$. Then, $\mathcal{D}^I_{a,b;c,d}(\boldsymbol{\rho}_n)$ for $a \leq b$ and $c \leq d$ is a formula that:
  - for $a < b$ and $c < d$ encodes that the weight represented by the sequences $\overline{d}_{a+1,n}, \ldots, \overline{d}_{b,n}$ and $\overline{d}_{c+1,n}, \ldots, \overline{d}_{d,n}$ belongs to the interval I,
  - for $a = b$ and $c < d$ encodes that the weight represented by the sequence $\overline{d}_{c+1,n}, \ldots, \overline{d}_{d,n}$ belongs to the interval I,
  - for $a < b$ and $c = d$ encodes that the weight represented by the sequence $\overline{d}_{a+1,n}, \ldots, \overline{d}_{b,n}$ belongs to the interval I,
  - for $a = b$ and $c = d$, the formula $\mathcal{D}^I_{a,b;c,d}(\boldsymbol{\rho}_n)$ is true iff $0 \in I$.

Thus, given the above, one can define the formula $[\widehat{\mathcal{M}}^{\varphi,\widehat{\iota}}]_k$ as follows:

$$[\widehat{\mathcal{M}}^{\varphi,\widehat{\iota}}]_k := \bigvee_{s \in \widehat{\iota}} I_s(\overline{\mathbf{w}}_{0,0}) \wedge \bigvee_{j=1}^{f_k(\varphi)} \overline{\mathbf{w}}_{0,0}$$

$$= \overline{\mathbf{w}}_{0,j} \wedge \bigwedge_{j=1}^{f_k(\varphi)} \bigwedge_{i=0}^{k-1} \mathcal{T}(\overline{\mathbf{w}}_{i,j}, ((\overline{a}_{i,j}, \overline{d}_{i,j}), \overline{\delta}_{i,j}), \overline{\mathbf{w}}_{i+1,j})$$

where $\overline{\mathbf{w}}_{i,j}$, $\overline{a}_{i,j}$, and $\overline{d}_{i,j}$ are, respectively, symbolic states, symbolic actions, and symbolic weights for $0 \leq i \leq k$ and $1 \leq j \leq f_k(\varphi)$. Hereafter, by $\boldsymbol{\rho}_j$ we denote the j-th symbolic k-run of the above unfolding, i.e., the sequence of transitions: $\overline{\mathbf{w}}_{0,j} \overset{(\overline{a}_{1,j}, \overline{d}_{1,j}), \overline{\delta}_{1,j}}{\longrightarrow} \overline{\mathbf{w}}_{1,j} \overset{(\overline{a}_{2,j}, \overline{d}_{2,j}), \overline{\delta}_{2,j}}{\longrightarrow} \ldots \overset{(\overline{a}_{k,j}, \overline{d}_{k,j}), \overline{\delta}_{k,j}}{\longrightarrow} \overline{\mathbf{w}}_{k,j}$.

The formula $[\varphi]_{\widehat{\mathcal{M}},k}$ encodes the bounded semantics of a WECTLK formula φ, and it is defined on the same sets of individual variables as the formula $[\widehat{\mathcal{M}}^{\varphi,\widehat{\iota}}]_k$. Moreover, it uses the auxiliary propositional formulae defined in [11].

Furthermore, following [9], our formula $[\varphi]_{\widehat{\mathcal{M}},k}$ uses the following auxiliary functions g_l, g_r, g_μ, $h_{\mathbf{U}}$, $h_{\mathbf{G}}$ that were introduced in [12], and which allow to divide the set $A \subseteq F_k(\varphi) = \{j \in \mathbb{N} \mid 1 \leq j \leq f_k(\varphi)\}$ into subsets needed for translating the subformulae of φ. Let $0 \leq n \leq f_k(\varphi)$, $m \leqslant k$, and $n' = min(A)$. The translation of WECTLK formula is defined in [9]. The theorem below states the correctness and the completeness of the presented SAT translation. It can be proved in a standard way, using induction on the complexity of the given WECTLK formula.

Theorem 2. *Let $\widehat{\mathcal{M}}$ be the abstract model for TWIS, and φ a WECTLK formula. For every $k \in \mathbb{N}$, $\widehat{\mathcal{M}} \models_k \varphi$ if, and only if, the propositional formula $[\widehat{\mathcal{M}}, \varphi]_k$ is satisfiable.*

Proof. The theorem can be proven by induction on the length of the formula φ (for details one can see [11]).

4 Experimental Results

In this section, we experimentally evaluate the performance of our SAT-based BMC encoding for WECTLK over the TWIS semantics. Note that, each multi-agent system, which can be modelled by TWIS, may be verified using presented module. The benchmark, we consider is the *timed weighted generic pipeline paradigm* TWIS abstract model (TWGPP) [13]. The abstract model of TWGPP involves $n+2$ agents: Producer producing data within the certain time interval ($[a,b]$) or being inactive, Consumer receiving data within the certain time interval ($[c,d]$) or being inactive within the certain time interval ($[g,h]$), a chain of n intermediate Nodes which can be ready for receiving data within the certain time interval ($[c,d]$), processing data within the certain time interval ($[e,f]$) or sending data. The weights are used to adjust the cost properties of Producer, Consumer, and of the intermediate Nodes. The precise description of the TWGPP system can be found in [13].

We assume the following two local weight functions for each agent:

- $d_P(Produce) = 4$, $d_P(send_1) = 2$, $d_C(Consume) = 4$, $d_C(send_{n+1}) = 2$, $d_{N_i}(send_i) = d_{Ni}(send_{i+1}) = d_{N_i}(Proc_i) = 2$,
- $d_P(Produce) = 4000000$, $d_P(send_1) = 2000000$, $d_C(Consume) = 4000000$, $d_C(send_{n+1}) = 2000000$, $d_{N_i}(send_i) = d_{Ni}(send_{i+1}) = d_{N_i}(Proc_i) = 2000000$.

The system is scaled according to the number of its Nodes (agents), i.e., the problem parameter n is the number of Nodes. For any natural number $n \geq 0$, let $D(n) = \{1, 3, \ldots, n-1, n+1\}$ for an even n, and $D(n) = \{2, 4, \ldots, n-1, n+$

1} for an odd n. Moreover, let $r(j) = d_P(Produce) + 2 \cdot \sum_{i=1}^{j} d_{N_i}(Send_i) + \sum_{i=1}^{j-1} \cdot d_{N_i}(proc_i)$.
Then, we define *Right* as follows: $Right = \sum_{j \in D(n)} r(j)$.

We consider the following formulae as specifications:

- $\varphi_1 = \mathbf{EF}_{[0,Right)}(ConsFree)$ - *it states that there exists a run on which Consumer receives a data and the cost of receiving the data will be less than Right.*
- $\varphi_2 = \mathbf{EF}_{[0,Right)}(ConsFree \wedge \mathbf{EG}(ProdSend \vee ConsFree))$ - *it states that there exists a run on which Consumer receives a data and the cost of receiving the data is less than Right and from that point there exists a run on which always either the Producer has sent a data or the Consumer has received a data.*
- $\varphi_3 = \overline{\mathrm{K}}_P(\mathbf{EF}_{[0,Right)}(ConsFree \wedge \mathbf{EG}(ProdSend \vee ConsFree)))$ - *it states that it is not true that Producer knows that there exists a run on which Consumer receives a data and the cost of receiving the data is less than Right and from that point there exists a run on which always either the Producer has sent a data or the Consumer has received a data.*
- $\varphi_4 = \overline{\mathrm{K}}_P(\mathbf{EF}_{[0,Right)}(ConsFree \wedge \overline{\mathrm{K}}_C \overline{\mathrm{K}}_P(\mathbf{EG}(ProdSend \vee ConsFree))))$ - *it states that it is not true that Producer knows that there exists a run on which Consumer receives a data and the cost of receiving the data is less than Right and at that point it is not true that Consumer knows that it is not true that Producer knows that there exists a run on which always either the Producer has sent a data or Consumer has received a data.*

The number of the considered k-runs is equal to 3 for φ_1, and 5 for φ_2, respectively. The length of the witness is $2 \cdot n + 4$ if $n \in \{1, 2\}$ and, $2 \cdot n + 2$ if $n > 2$ for the formula φ_1, $2 \cdot n + 2$ for the formula φ_2, respectively.

Performance evaluation. We performed our experimental results on a computer equipped with I7-5500U processor, 12 GB of RAM, and the operating system Ubuntu Linux with the kernel 4.4.0. Our SAT-based BMC algorithm was implemented as a standalone program written in the programming language `C++`. We used the state of the art SAT-solvers: CryptoMiniSAT https://github.com/msoos/cryptominisat and Yices (SAT version) [2] (http://yices.csl.sri.com/), and state of the art SMT-solvers: Z3 [6] (for the old and the new one versions of SMT-BMC implementation), and Yices (SMT version) [2].

For both properties φ_1 and φ_2 we scaled up both the number of nodes and the weights parameters. The results are summarised on charts in Fig. 1. For the formula φ_1 memory usage for the SAT-BMC is very high. In this case the new SAT-BMC implementation with Yices-SAT can verify only 25 nodes for the basic weights (`bw` for short) and 23 nodes for `bw` multiplied by 10^6. CryptoMiniSAT can verify only 23 nodes for `bw` and `bw` multiplied by 10^6. One can observe that our improved the SMT-based BMC is better than the old one, and it is not sensitive to scaling up the weights, but it is sensitive to scaling up the size of benchmark. For `bw` the SMT-BMC based on Z3 can verify 29 nodes, and the SMT-BMC based on Yices-SMT can verify 27 nodes. For `bw` multiplied by 10^6

Z3 is able to verify 29 nodes, and Yices-SMT 27 nodes, however total memory usage is better for Yices-SMT. In the case of the formula φ_2 SAT-based BMC is able to verify 23 nodes for `bw` and CryptoMiniSAT, and 21 nodes for Yices-SAT; for `bw` multiplied by 10^6 CMS is able to verify 21 nodes, and Yices-SAT is able to verify 19 nodes, however memory usage for SAT-BMC is very high. In the case of the SMT-based approach Z3 is able to verify only 15 nodes for `bw`, and 19 nodes for `bw` multiplied by 10^6 and Yices-SMT. As we can see on charts in Fig. 1 Yices-SMT has the lowest memory usage. The old SMT-BMC approach is he worst one.

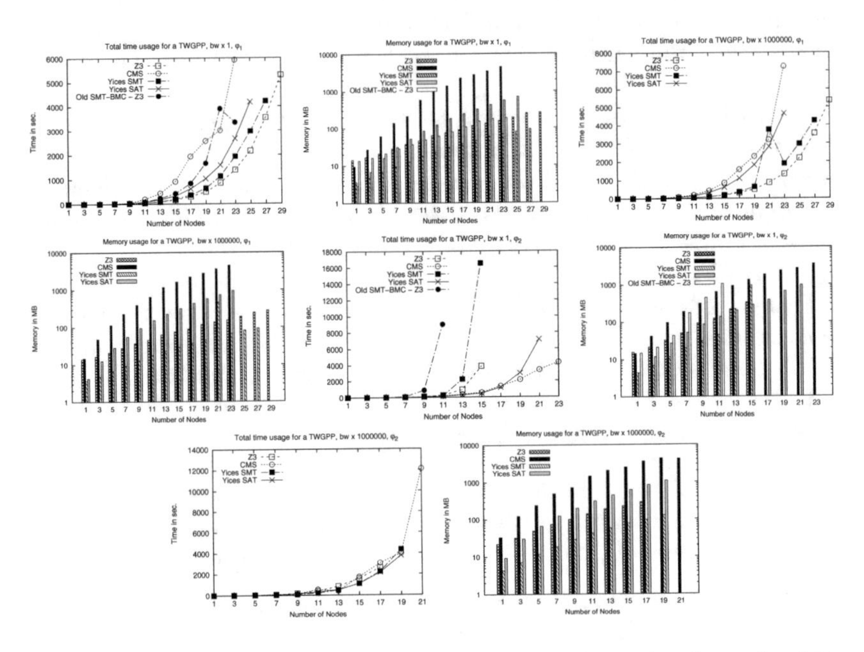

Fig. 1. Formulae: φ_1 and φ_2: Scaling up both the number of nodes and weights.

The results for the properties φ_3 and φ_4 are summarised on charts in Fig. 2. For the formula φ_3 the SMT-BMC approach (the old one and the new one) is much better than SAT-BMC approach. Z3 was able to verify 29 nodes in both cases. The SAT-BMC with Yices-SAT was able to verify only 25 nodes for `bw`, and 23 nodes for `bw` multiplied by 10^6. Also the total memory usage for SAT-BMC is much worse than for SMT-based approach. In the case of the property φ_4 SAT-based BMC with CryptoMiniSAT is much better. It is able to verify 23 nodes for `bw`, and 21 nodes for `bw` multiplied by 10^6, however total memory usage is very high. Also in this case the old SMT-based implementation is the worst one. In the comparison of SMT-BMC approaches the best one is Z3 for `bw`, and Yices-SMT for `bw` multiplied by 10^6.

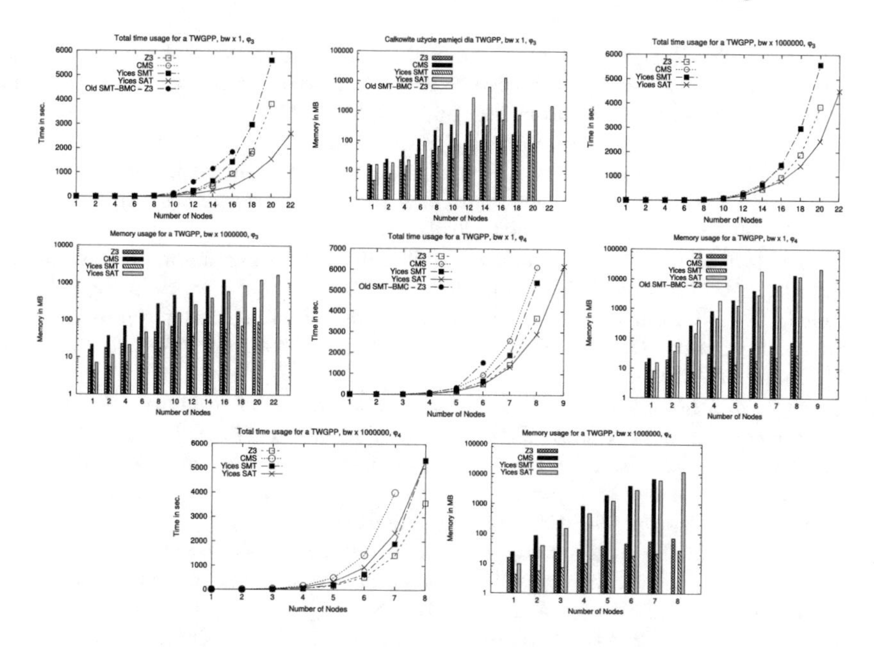

Fig. 2. Formulae: φ_3 and φ_4: Scaling up both the number of nodes and weights.

5 Conclusions

We have proposed, implemented, and experimentally evaluated SAT-based bounded model checking approach for WECTLK interpreted over the timed weighted interpreted systems. We compared the SAT-BMC approach and the SMT-BMC approach for different kinds of SAT- and SMT-solvers. We also improved the old implementation of the SMT-BMC [13] in terms of memory usage. The experimental results show that the approaches are complementary. They show that the choice of the BMC method and SAT- or SMT-solver should depend on the considered formula. The module will be added to the model checker VerICS ([4]). All the benchmarks together with an instruction how to reproduce our results can be found at the webpage http://tinyurl.com/bmc-twis-wectlk.

References

1. Clarke, E.M., Grumberg, O., Peled, D.A.: Model Checking. The MIT Press, Cambridge (1999)
2. Dutertre, B.: Yices 2.2. In: Computer Aided Verification - 26th International Conference, CAV 2014, Held as Part of the Vienna Summer of Logic, VSL 2014, Vienna, Austria, 18–22 July 2014, Proceedings, pp. 737–744 (2014)
3. Fagin, R., Halpern, J.Y., Moses, Y., Vardi, M.Y.: Reasoning About Knowledge. MIT Press, Cambridge (1995)

4. Kacprzak, M., Nabialek, W., Niewiadomski, A., Penczek, W., Pólrola, A., Szreter, M., Woźna, B., Zbrzezny, A.: VerICS 2007 - a model checker for knowledge and real-time. Fundamenta Informaticae **85**(1–4), 313–328 (2008)
5. Męski, A., Penczek, W., Szreter, M., Woźna-Szcześniak, B., Zbrzezny, A.: BDD-versus SAT-based bounded model checking for the existential fragment of linear temporal logic with knowledge: algorithms and their performance. Auton. Agents Multi-Agent Syst. **28**(4), 558–604 (2014)
6. Moura, L., Bjørner, N.: Z3: an efficient SMT solver. In: Ramakrishnan, C.R., Rehof, J. (eds.) TACAS 2008. LNCS, vol. 4963, pp. 337–340. Springer, Heidelberg (2008)
7. Peled, D.: All from one, one for all: on model checking using representatives. In: Courcoubetis, C. (ed.) CAV 1993. LNCS, vol. 697, pp. 409–423. Springer, Heidelberg (1993)
8. Penczek, W., Lomuscio, A.: Verifying epistemic properties of multi-agent systems via bounded model checking. Fundamenta Informaticae **55**(2), 167–185 (2003)
9. Woźna-Szcześniak, B.: SAT-based bounded model checking for weighted deontic interpreted systems. In: Correia, L., Reis, L.P., Cascalho, J. (eds.) EPIA 2013. LNCS, vol. 8154, pp. 444–455. Springer, Heidelberg (2013)
10. Woźna-Szcześniak, B., Zbrzezny, A.M., Zbrzezny, A.: SAT-based bounded model checking for weighted interpreted systems and weighted linear temporal logic. In: Boella, G., Elkind, E., Savarimuthu, B.T.R., Dignum, F., Purvis, M.K. (eds.) PRIMA 2013. LNCS, vol. 8291, pp. 355–371. Springer, Heidelberg (2013)
11. Wozna-Szczesniak, B.: SAT-based bounded model checking for weighted deontic interpreted systems. Fundam. Inform. **143**(1–2), 173–205 (2016)
12. Zbrzezny, A.: Improving the translation from ECTL to SAT. Fundamenta Informaticae **85**(1–4), 513–531 (2008)
13. Zbrzezny, A.M., Zbrzezny, A.: Checking WECTLK properties of timed real-weighted interpreted systems via SMT-based bounded model checking. In: Pereira, F., Machado, P., Costa, E., Cardoso, A. (eds.) EPIA 2015. LNCS, vol. 9273, pp. 638–650. Springer, Cham (2015)

Agent-Based Modeling and Simulation

Communication and Autonomous Control of Multi-UAV System in Disaster Response Tasks

Maher Aljehani(✉) and Masahiro Inoue

Shibaura Institute of Technology, Graduate School of Engineering and Science, Saitama, Japan
nb16507@shibaura-it.ac.jp, inouem@sic.shibaura-it.ac.jp

Abstract. After disasters occurrence, advanced technologies always play significant contributions to various disaster response tasks. For instance, utilization of Unmanned Aerial Vehicle (UAV) for aerial imagery data helps to get a real-time monitoring system of the stricken areas. Recently, UAV is becoming a ubiquitous system and valuable technology in many civil applications, motivating researchers to endeavor to develop UAVs systems. Moreover, Internet of Things (IoT) has been employed to several study cases of disaster response. Using IoT to communicate, manage and control multiple UAVs after disaster occurrence is a real practical approach. However, this kind of integration wasn't clearly presented especially in the technical sides like communication and autonomous control. Consequently, this paper aims to reveal the scientific scenes of communications and controls between multiple UAVs and Ground Control Stations (GCSs). Due to an unexpected failure within a single UAV system, this study presents a multi-UAV system for scanning and tracking missions. Authors turned UAV to an IoT device by using embedded LTE dongle on the UAV control board. In the beginning of the paper, authors raised the issues of existing systems. Then, presented the design and process of the proposed system. Lastly, they demonstrated some results through experiments.

Keywords: Unmanned aerial vehicles · Communication · Autonomous control · TCP · UDP · MAVlink · IoT · Disaster responses

1 Introduction

The aerial visual data of buildings, roads, and pedestrian traffics after disaster occurrence are very important information to create a new safe map [1]. Furthermore, providing real-time images can assist to pinpoint the exact geographical locations of refugees and victims before planning to guide them to safe locations [2]. Therefore, employing a Multi-UAV system with integration of IoT for disaster response is very appreciated, especially for rescue operations [3].

G. Jezic et al. (eds.), *Agent and Multi-Agent Systems: Technology and Applications*, Smart Innovation, Systems and Technologies 74,
DOI 10.1007/978-3-319-59394-4_12

Authors in the previous study used a single UAV to track pedestrians in a scenario of after disaster occurrence [4]. However, in a case of multiple missions, using one single UAV is considered as a curtailment of the system functionality; since it has many flaws in various features comparing to Multi-UAV system [1].

IoT is a suitable solution for autonomous missions and controlling multi-UAV system from one GCS. Normal controller (i.e. radio controller) has implementation issues in controlling Multi-UAV system. For instance, designation of multiple operators is a necessary procedure in order to control multiple UAVs simultaneously; furthermore, it has a short-range control problem. However, in a case of cellular network coverage or multiple Internet Access Points (AP), problems of short-range control and multiple operators designation aren't flaws anymore. Of course, after disaster occurrence, cellular towers may collapse. In this point, emergency communication can be implemented like mobile satellite Internet systems or Mobile Ad-hoc Network (MANET) which basically works as a temporary network provider.

2 Communication

Employing UAV as an IoT device with embedded sensors and cameras makes UAV instructable for various of missions through the Internet. To initiate a certain required, the Internet AP has to be utilized in the system. The Internet AP also can be used to communicate with refugees, like sending alert messages and guiding them by using a new safe map. In the meanwhile, it makes UAV accessible agent in the network. For attaining a dynamic communication, UAV acts as a server in remote and nearby GCS. The communication in nearby GCS is going to be through ZigBee Network or 3DR telemetry radio. However, in a case of using remote GCS, the communication is going to be through cellular coverage or Internet AP as Fig. 1 demonstrates. Firstly, UAV is going to receive commands to coordinate itself. Then, it sends images back to the GCS. The location of the operators isn't a flaw in this system since it used a long range communication and wide internet connection coverage (e.g. cellular coverages for remote GCS).

2.1 Internet of Things

Elastic Compute Cloud (EC2) in AWSIoT established an ArduPilotMega (APM) connection [5] that mainly used MAVlink protocol [8] to connect to UAVs via UDP/TCP connection. The connection between UAV and GCS can be initiated through ZigBee in a case of nearby GCS mode or LTE for remote GCS as we mentioned earlier. In either case, IP addresses of the UAVs and GCS must be acknowledged. In LTE/3G telemetry data connection, the problem of assigning new arbitrary IP addresses and dynamic ISP in the network (i.e. DHCP) can be solved by implementing a Dynamic DNS (DynDNS) or configuring a Virtual Proxy Network server (VPN) [6]. After forwarding the embedded dongle modem, The accessibility can be done via remote protocols like SSH on port 22 protocol

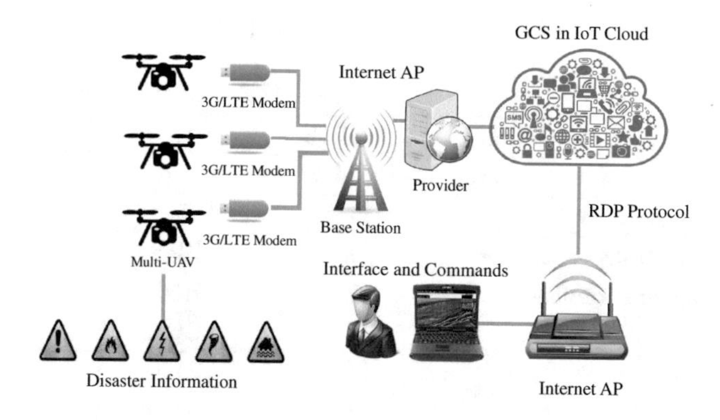

Fig. 1. Network system of Multi-UAV using internet AP

or RDP on port 3389. Also, In VPN mode, tunnels between clouds and the UAVs provide a secure end-to-end communication as the Fig. 2 shows. In the beginning, operators need to setup dynamic networks for both UAV and EC2 that has GCS. Also, gateways need to be comprehensively configured.

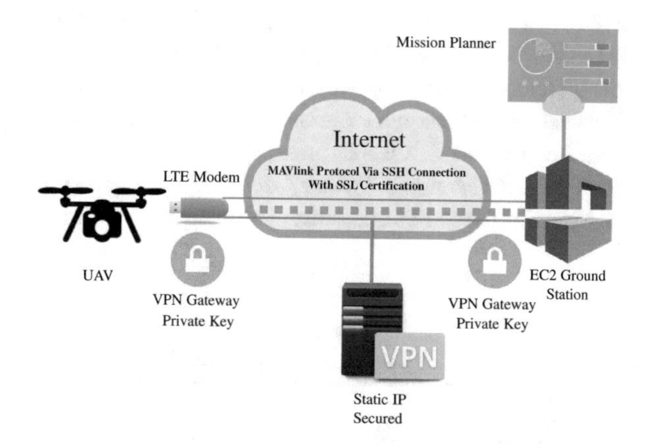

Fig. 2. Communication between each UAV and EC2 of IoT using VPN

2.2 Multi-UAV Communication System

In the early usage of UAV system, it only used one huge single UAV for executing one or multiple tasks. Single UAV system is not sufficient comparing to Multi-UAV system, even if it was an advanced UAV that has high specifications. During the disaster, the timeline is important to be short to save what can be saved in such incident. Therefore, multi-UAV system can collaboratively finish multi-assignments more objectively rather than a single UAV system, especially in the time of finishing missions. Table 1 illustrates the different characteristics and features between single UAV system and multi-UAV system [7].

Table 1. Multi-UAV and single UAV systems.

Features	Single UAV	Micro multi-UAV
Failure of completion	High	Low
Speed of finishing assignments	Low	High
Survivability	Poor	High
Multitasks capability	Low	High
Reconfigurability	Low	High
Ad-Hoc networks	Cannot	Optimal
Complexity	Low	High
Heterogeneous	Inapplicable	Applicable

2.3 MAVlink Protocol

MAVlink is an acronym of Micro Air Vehicle Link [8]. It is a protocol that helps UAV to communicate and interact with GCS. MAVlink protocols can be defined as a large number of waypoints command types that can be sent wirelessly to the UAV flight controller. To retrieve a list of all waypoints from GCS, a WAYPOINT_REQUEST_LIST message has to be sent by UAV first. Then, GCS will response with a WAYPOINT_COUNT messages stating the number of waypoints list. After that, UAV will ask for all waypoints starting with a sequence number of 0 and that can be accomplished by sending WAYPOINT_REQUEST message. As soon as GCS received the request message, it has to answer to every request with a corresponding of WAYPOINT messages. When the last waypoint has successfully retrieved, UAV sends a WAYPOINT_ACK message to the GCS and execute the mission. Figure 3 demonstrates the messages of waypoints between GCS and UAV.

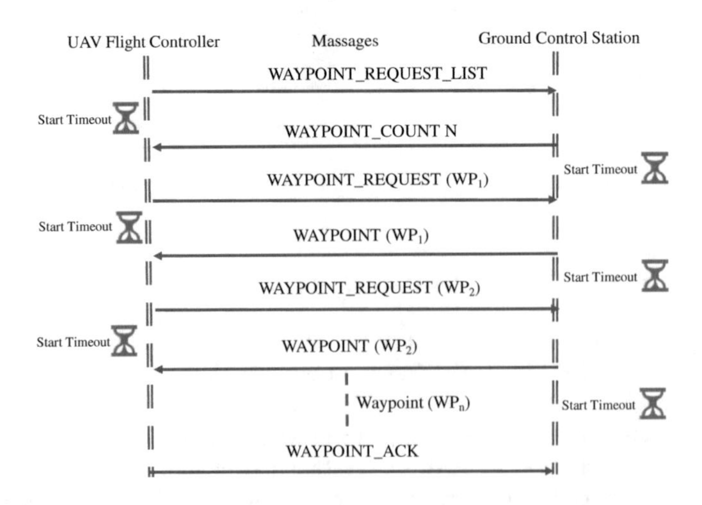

Fig. 3. MAVlink messages of waypoints between GCS and UAV

3 Control System

3.1 Tracking Mission

Tracking mission is mainly used to recorder routes history and trajectories of refugees in the impacted areas. After evaluation, cloud generates a map according to the refugee's paths. In the work of [4], UAV managed to track human

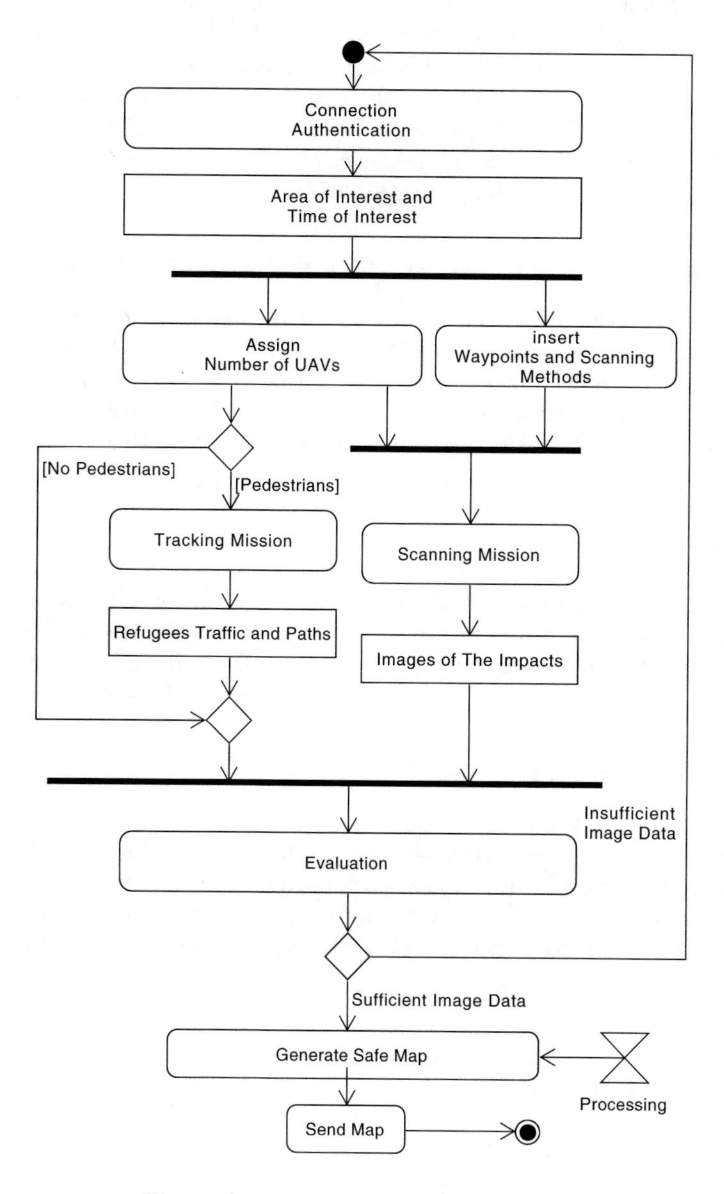

Fig. 4. Activity diagram of UAV missions

successfully by using image processing and send GPS data of the UAV to the cloud as safe routes. EC2 has a GCS software that can assign UAVs to track refugees and record their directions. In that study, authors used tracking histograms of oriented gradients (HOG) for human detection and tracking [9].

3.2 Scanning Mission

Scanning mission is different than tracking mission in terms of control and process. In tracking mission, waypoints of the UAV depend on the movements of targets (i.e. humans, bicycles, and cars). However, in scanning mission, the waypoints are already preprogrammed. Scanning mission can be also defined as a survey mission. Basically, after disaster occurrence, UAV will survey the impacted area and send the aerial images to the cloud to generate an emergency map. Figure 4 shows the activity diagram of the scanning and tracking tasks.

3.3 Autonomous Control of Tracking and Scanning Missions

Recently, controlling multi-UAV system has attracted a lot of attentions in robotics communities. Autonomous control of the UAVs can be done by many methods [10]. Figure 5 shows the autonomous control that have been used in both missions. In this study, scanning task used a Sense-Plan-Act control system. On the other hand, in tracking task the control method was Reactive-Robotic control system.

3.4 UAV Model

The UAV model in this study was a hexacopter type. Generally, copter's frames have the ability to do hovering mode which helps to get more stable imaging angles. Furthermore, by using multi-rotor types, there is no need to maintain velocity to avoid a crash or falling down like Fixed wing frames.

3.5 Remote and Nearby GCS

The location of the controller is not an issue since the system has to be dynamic and easy to reconfigure according to the requirements of the manual interruptions. So, in remote GCS, the methodology of controlling is different than nearby GCS as Fig. 6 shows. In this system, both missions are autonomous missions; however, manual control is necessary as an emergency control interruption.

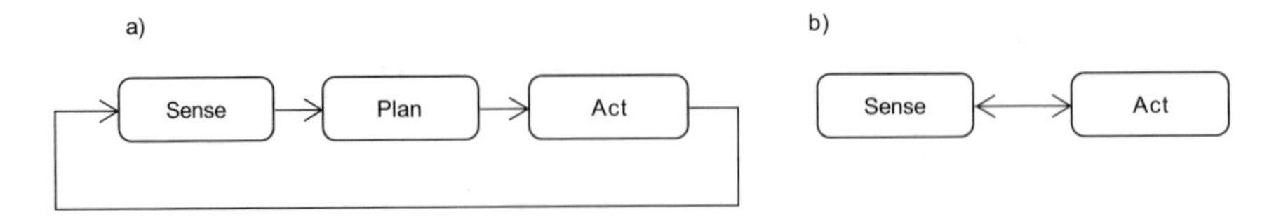

Fig. 5. (a) Scanning control (b) Tracking control

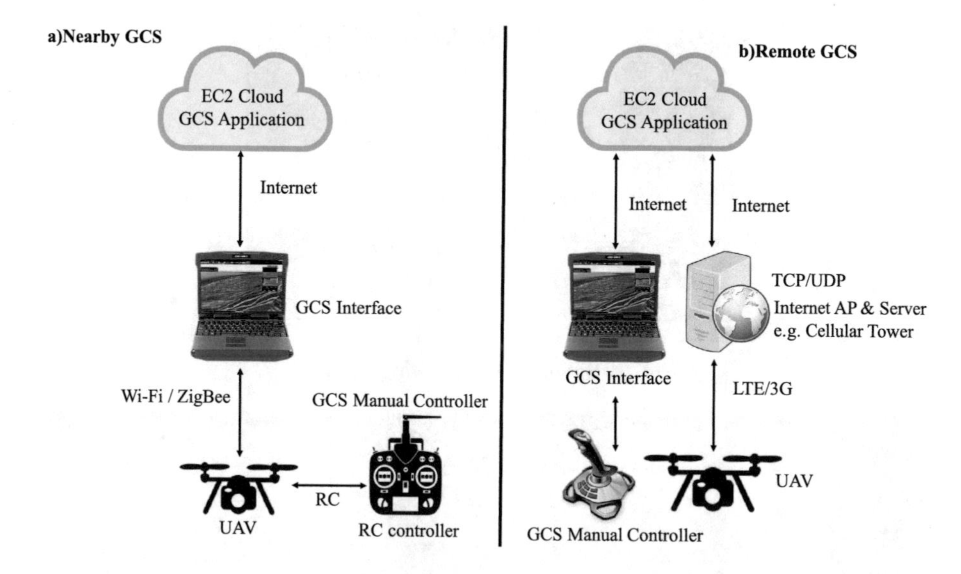

Fig. 6. (a) Nearby GCS (b) Remote GCS

4 Experiments

The GCS in this system is an open source software called "Mission Planner", it has been installed in Windows OS 2016 on EC2 cloud server. UAV flight controller is powered by Raspberry Pi's HAT control shield named Navio2 [11]. Since communications between UAVs and EC2 computing clouds have to be evaluated, we monitored UDP and TCP in real-time by software called Wireshark. The goal of these experiments is to evaluate the network performance of EC2 cloud through MAVlink protocol and LTE/3G dongle gateway. After finishing some network configurations, we connected UAVs to GCS (i.e. Mission Planner on EC2) to analyze TCP and UDP packet traffic per second during the missions.

4.1 Mission Planning

In this experiment, we used a single UAV mission planning as Fig. 7 shows. We wrote a course with many waypoints into the autopilot of UAV's flight controller. The experiments were on the university campus that has 0.16 km^2 area of interest. The mission took around 16 min and that didn't include the taking off and landing time. However, 16 min range is an inadequate performance for providing disaster information and to send a proper information of the emergency map. Also, factors like altitude, way of scanning and speed are very important pillars for the mission time cost.

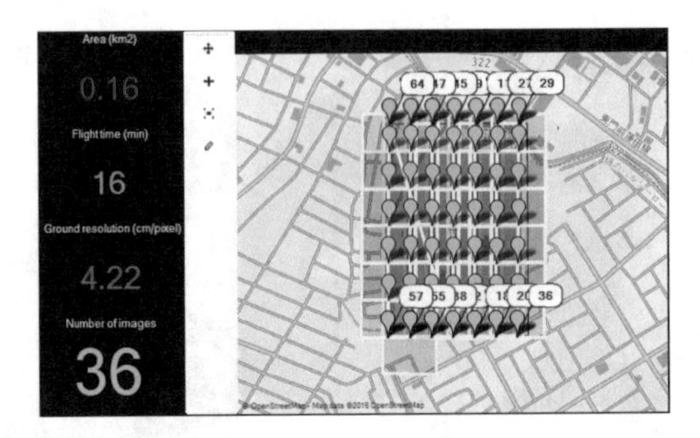

Fig. 7. Single UAV scanning mission

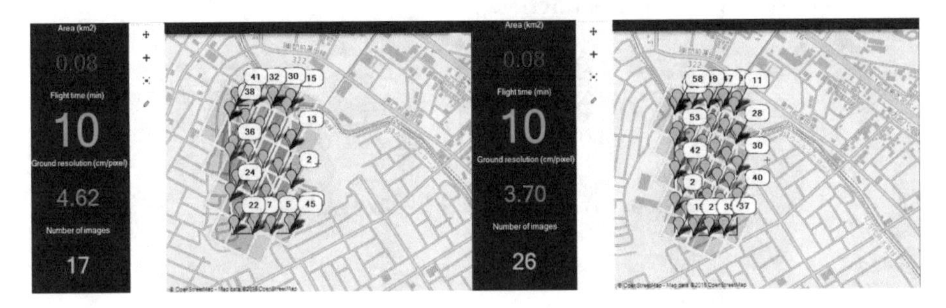

Fig. 8. Example of two instances of EC2 executing scanning mission

4.2 Multi-UAV Scanning

As Fig. 8 shows, in the same area of interest, we assigned two scanning UAVs. Practically, assignment of two UAVs is appropriate for 10 min time of interest. For more demonstration, Formula 1 presents an equation that simplified the relationship between time of interest, area of interest and the number of UAVs. Theoretically, the less time of interest wanted, the more number of UAVs has to be used. Of course, using two UAVs in the same area of interest decreased the time of the mission. However, besides the less time of interest, Multi-UAV enhanced the quality of emergency data since the images of all orientations at the stricken areas have been provided at the same time (Table 2).

- Altitude = 150 m
- Speed = 5 m/s
- Area of Interest = 0.16 km^2

$$\therefore T_{scan} = \frac{AreaOfInterest(km^2)}{(NumberOfUAVs) * (SingleUAV_{scan}(km^2/min))} = min \tag{1}$$

4.3 MAVlink Protocol over UDP and TCP

In order to evaluate the network performance, we analyzed the behaviors of UDP and TCP connections over LTE and MAVlink protocol during missions. MAVlink

Table 2. Two UAVs and single UAV in scanning mission

Parameter	Single UAV	Two UAVs
Area (km^2)	0.16	0.16
Mission flight time (min)	15:57	10:05
Distance (km)	3.83	4.53
Number of images	36	43
Distance between lines (m)	50.64 m	55.44
Footprints (m)	168.8×126.6	352.6×279.5

protocol supports UDP and TCP protocols connection to send telemetry data and receive commands from the GCS. In this experiment, the signal power was between −110 dBm and −79 dBm and it was diversely changing. We ran missions in the EC2 using MAVlink protocols through TCP and UDP. Then, we recorded the pockets traffic per second between GCS and UAV as Fig. 9 shows.

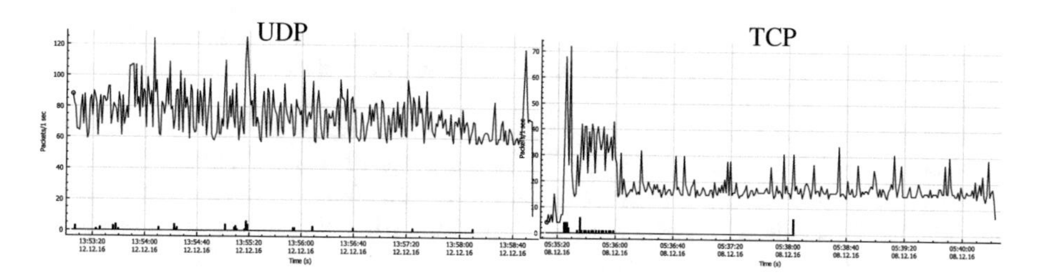

Fig. 9. Graph of TCP and UDP PacketsTraffic per second

4.4 Discussion

Regarding the experiment results, UDP and TCP have different characteristics. Therefore, they behaved quite differently during the missions. Theoretically, TCP has many advantages over UDP protocols. For example, TCP guarantees the data transmission to be received and manages the sequence of the data without depublication. However, TCP is a heavy protocol and doesn't have a feature of broadcasting. Alternatively, UDP can work as broadcast protocol and it's lighter than TCP. In the experiment, UDP used more packet than TCP and that because TCP connection in MAVlink protocol used some UDP protocols to transfer telemetry data to GCS. On the other hand, in UDP connection, MAVlink protocol used UDP only without any TCP. Therefore, UDP looks had more traffic than TCP.

5 Conclusion

This study shows how Multi-UAV system contribution can execute multiple rescue tasks like scanning and tracking autonomously and simultaneously using IoT as GCS. Also, it demonstrates a combination of tracking data and scanning mission which can create a high-quality safe map for the refugees after the disaster occurrence. The footprints of two UAVs is more than single UAV with less time and that proved the reliability of the Multi-UAV system over single UAV system. In this study, UAV is a server for EC2 and EC2 is an interface for operator. Using DynDNS or VPN gave a secure connection between UAVs and the GCS in EC2 and also solved the issues of random IP assignment. The experiments revealed that MAVlink cannot establish a TCP connection without UDP protocol. The next step in this research is to keep improving the communication by using Multi-hop communication system with leader UAV at simulated disaster area based on real scenarios.

References

1. Aljehani, M., Inoue, M.: Multi-UAV tracking and scanning systems in M2M communication for disaster response. In: 2016 IEEE 5th Global Conference Consumer Electronics, pp. 1–2 (2016)
2. Doherty, P., Rudol, P.: A UAV search and rescue scenario with human body detection and geolocalization. In: Orgun, M.A., Thornton, J. (eds.) AI 2007. LNCS, vol. 4830, pp. 1–13. Springer, Heidelberg (2007). doi:10.1007/978-3-540-76928-6_1
3. Sakano, T., Fadlullah, Z.M., Ngo, T., Nishiyama, H., Nakazawa, M., Adachi, F., Kato, N., Takahara, A., Kumagai, T., Kasahara, H., Kurihara, S.: Disaster-resilient networking: a new vision based on movable and deployable resource units. IEEE Netw. **27**(4), 40–46 (2016)
4. Aljehani, M., Inoue, M.: Generating evacuation routes by using drone system and image analysis to track pedestrian and scan the area after disaster occurrence. In: The 10th SEATUC (2016). doi:10.13140/RG.2.1.4749.9121
5. ArduPilot Mega APM. http://diydrones.com/notes/ArduPilot
6. Coonjah, I., Catherine, P.C., Soyjaudah, K.M.S.: Experimental performance comparison between TCP vs UDP tunnel using OpenVPN. In: IEEE the (ICCCS) Computing, Communication and Security International Conference, pp. 1–5 (2015)
7. Gupta, L., Jain, R., Vaszkun, G.: Survey of important issues in UAV communication networks. IEEE Commun. Surv. Tutorials **18**(2), 1123–1152 (2015)
8. QGroundControl.: MAVlink Micro Air Vehicle Communication Protocol - QGroundControl GCS. http://qgroundcontrol.org/MAVlink/start
9. Dalal, N., Triggs, B.: Histograms of oriented gradients for human detection. In: (CVPR2005) IEEE Computer Society Conference on Computer Vision and Pattern Recognition, pp. 886–893 (2005)
10. Krzysztof, W., Pawea, L.: Proposed algorithms for mission planning for groups of UAVs. In: European Seventh Framework Programme FP7-218086-Collaborative Project (2010)
11. Emild Navio2, A Raspberry Pi's HAT Autopilot Shield. https://emlid.com/introducing-navio2/

Decision Function Implementation in MAREA Simulations Influencing Financial Balance of Small-Sized Enterprise

Roman Šperka(✉) and Dominik Musil

Department of Business Economics and Management,
School of Business Administration in Karviná, Silesian University in Opava,
Univerzitní nám. 1934/3, 733 40 Karviná, Czech Republic
{sperka,0150860}@opf.slu.cz

Abstract. The aim of this paper is to present the use of a decision function in the implementation of a multi-agent simulation model of a small-sized enterprise dealing with trading. The subject of the presented research are simulation experiments in MAREA software framework, which was designed to simulate trading behaviour of a trading company. Firstly, we present a multi-agent system and a mathematical description of a decision function, which is used to establish the price of traded goods. Secondly, we present MAREA software framework and lastly we discuss the simulation results of company dealing with retailing of fluorescence colours. The results obtained show that simulation experiments in MAREA could be used to support the decision-making process of a management of trading companies in the scope of predicting key performance indicators and changes of parameters and their impact on the company's financial balance.

Keywords: Multi-agent system · Framework · Model · Simulation · Software · Business process · Trading · MAREA

1 Introduction

The importance of business information systems has been rapidly growing recently because of the globalization. The managements of business companies have to increase the flexibility and the decision speed to keep pace with the situation on targeting markets. The complexity of business operations often does not allow to take measures without known impacts of such decisions. This is where the modeling and simulations find their place (e.g., [1]). While analytical modeling approaches are based mostly on mathematical theories [2, 3] the approach used in this paper is based on simulations.

The simulations, we experiment with, can be described as agent-based [4] in the field of business economics. In our opinion only several problems can be identified while using classical simulation approaches (e.g., [5]). There is a lot of other influences that cannot be captured by using typical business process models (e.g., the effects of collaboration of business process participants or their communication, experience level, cultural or social factors, etc.) as shown in, e.g., [6].

G. Jezic et al. (eds.), *Agent and Multi-Agent Systems: Technology and Applications*, Smart Innovation, Systems and Technologies 74,
DOI 10.1007/978-3-319-59394-4_13

Intelligent software agents representing business process participants are more accordant with people and can model a typical behaviour like communication, coordination or cooperation – the basic characteristics of a multi-agent system (MAS). Software agents can also be specialized (e.g., adaptability in a new environment or in life experience). They are able to plan the tasks and to assign the work to other agents. Intelligence of a MAS is created emergently during the interaction both among the agents themselves, with their environment, and its components.

The presented research is based on the decision function in a control loop model [7, 8] of a generic business company. The control loop consists of controlled units like sales, purchase, production and others, managed by a regulator unit (the management of the company). The outputs of the controlled units are measured by the measuring unit and compared with the desired key performance indicators (KPIs). The differences found are sent to the regulator unit, which takes the necessary measures in order to keep the system in the closeness to the KPI values. However, it was shown that a business company must be looked upon as a system with social functions and responsibilities, where individuals besides the company KPIs also follow their personal aims and preferences (e.g., the paper from [9], summarizing the Corporate Social Responsibility research of many other authors). The same can be observed in the market, where the customers and the suppliers follow their own targets. The principles described could be used to improve decision making processes of the company's management.

Previous research results of our approach to this challenge using software agents were presented in [8, 10–13]. Business process simulation framework called MAREA was implemented and described recently in [14].

The structure of the paper is as follows. In the second section the business process simulation and the mathematical model are described. In the third and last section MAREA software framework is introduced and simulation results are discussed.

2 Multi-agent System Details

To ensure the outputs of trading processes simulations a simulation framework was implemented and used to trigger the simulation experiments. The framework covers business processes supporting the selling of goods by company sales representatives to the customers – seller-to-customer negotiation (Fig. 1). It consists of the following types of agents: sales representative agents (representing sellers, seller agents), customer agents, an informative agent (provides information about the company market share, and company volume), manager agent (manages the seller agents, calculates KPIs), and disturbance agent (brings disturbances of market environment into the model). All the agent types are developed according to the multi-agent approach. The interaction between agents is based on the FIPA contract-net protocol [15].

The number of customer agents is significantly higher than the number of seller agents in the model because the reality of the market is the same. The behavior of agents is influenced by two randomly generated parameters using the normal distribution (an amount of requested goods and a sellers' ability to sell the goods). In the lack of real information about the business company, there is a possibility to randomly generate different parameters (e.g. company market shares for the product, market volume for the product in local currency, or a quality parameter of the seller). The influence of randomly generated parameters on the simulation outputs while using different types of distributions was previously described in [11].

In the text to follow, the seller-to-customer negotiation workflow is described and the mathematical definition of a decision function is proposed. Decision function is used during the contracting phase of agents' interaction. It serves to set up the limit price of the customer agent as an internal private parameter. One stock item simplification is used in the implementation. Participants of the contracting process in our multi-agent system are represented by the software agents - the seller and customer agents interacting in the course of the quotation, negotiation and contracting. There is an interaction between them. The behavior of the customer agent is characterized in our case by a decision function (Eq. 1).

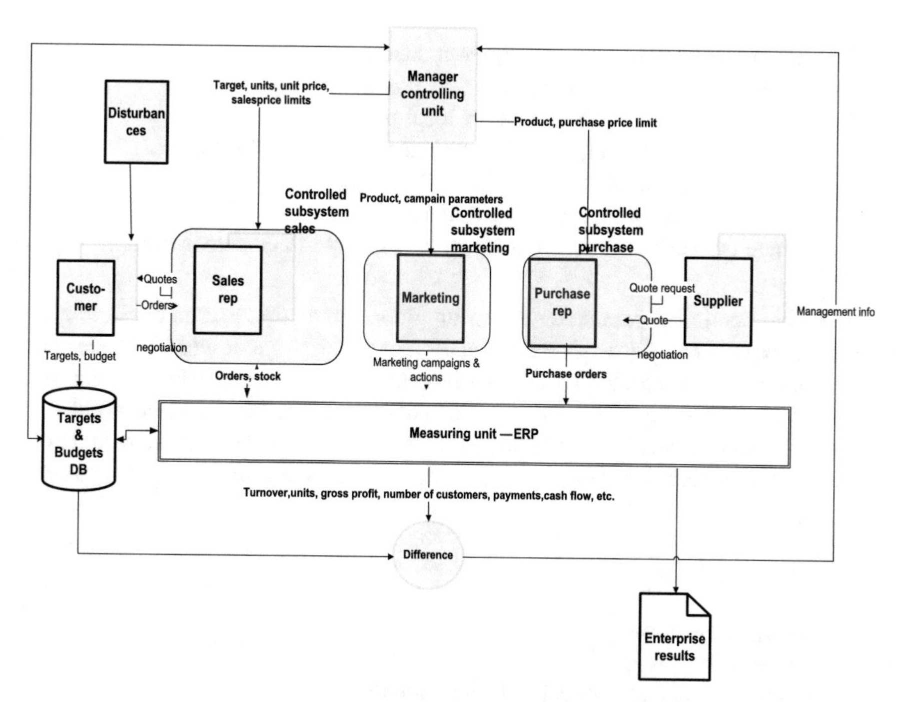

Fig. 1. Generic model of a business company. (Source: adapted from [14])

Each period turn (here we assume a week), the customer agent decides whether to buy something. His decision is defined randomly. If the customer agent decides not to buy anything, his turn is over; otherwise he creates a sales request and sends it to his seller agent. The seller agent answers with a proposal message (a certain quote starting with his maximal price: (*limit price * 1.25*). This quote can be accepted by the customer agent or not. The customer agents evaluate the quotes according to the decision function. The decision function was proposed to reflect the company's market share for the product quoted (a market share parameter), seller's ability to negotiate, total market volume for the product quoted etc. (in e.g., [11]). If the price quoted is lower than the customer's price obtained as a result of the decision function, the quote is accepted. In the opposite case, the customer rejects the quote and a negotiation is started. The seller agent decreases the price to the average of the minimal limit price and the current price (in every iteration is getting effectively closer and closer to the minimal limit price), and resends the quote back to the customer. The message exchange repeats until there is an agreement or a reserved time passes.

The decision function for the m-th seller pertaining to the i-th customer determines the price that i-th customer accepts (adjusted according to [11]).

$$c_n^m = \frac{\tau_n T_n \gamma \rho_m}{O v_n} \quad (1)$$

c_n^m - price of n-th product offered by m-th seller,
τ_n - market share of the company for n-th product $0 < \tau_n < 1$,
T_n - market volume for n-th product in local currency,
γ - competition coefficient, lowering the success of the sale $0 < \gamma < 1$,
ρ_m - m-th sales representative ability to sell $0,5 \leq \rho_m \leq 2$,
O - number of sales orders for the simulated time,
v_n - average quantity of the n-th product, ordered by i-th customer from m-th seller.

The aforementioned parameters represent global simulation parameters set for each simulation experiment. Other global simulation parameters are: lower limit sales price, number of customers, number of sales representatives, number of iterations, and mean sales request probability. The more exact parameters can be delivered by the real company, the more realistic simulation results can be obtained. In case we would not be able to use the expected number of sales orders O following formula can be used

$$O = ZIp \quad (2)$$

Z - number of customers,
I - number of iterations,
p - mean sales request probability in one iteration.

Customer agents are organized in groups and each group is being served by concrete seller agent. Their relationship is given; none of them can change the counterpart. Seller agent is responsible to the manager agent. Each turn, the manager agent gathers data from all seller agents and stores KPIs of the company. The data is the result of the simulation and serves to understand the company behavior in a time – depending on the agents' decisions and behavior. The customer agents need to know some information about the market. This information is given by the informative agent. This agent is also responsible for the turn management and represents outside or controllable phenomena from the agents' perspective.

3 Software Framework Description and Simulation Results

In this section, an enhanced software prototype of a framework MAREA based on agent-based trading company control loop and simulation results are introduced. The prototype is based on the research results presented herein above. The Enterprise Resource Planning system (ERP) using the REA ontology approach is used as a measuring and storing element in the framework and is a part of it. The system has been developed in cooperation between Silesian University in Opava, School of Business Administration in Karvina, Czech Republic and REA technology Copenhagen, Denmark. After the prototype tests at the end of the year 2011, it was presented at the beginning of 2012 for the first time [16].

3.1 MAREA Description

Framework enables users to set up trading company parameters and run trading simulation for a specific time to interpret the development of KPIs of the company. It consists of two main components, the Simulation of a multi-agent system (MAS) and the ERP system. A simulation designer can either use the ERP system directly, or can program intelligent agents to perform the same activities that a human user can perform. For example, a simulation designer can use the ERP system directly to create initial data for a simulation, then start the agent platform to run a simulation, then using the ERP system inspect the simulation results, and even adjust the data (within the rules implemented in the ERP system) and then start the agent platform to continue running the simulation. Both agents and a human user can read data from the ERP system, write data to the ERP system, and perform actions, such as sending a purchase order.

Simulation of negotiation between agents about sales and purchases is one of the key functions of the multi-agent simulation system. The messages the agents send to each other during negotiation are recorded in the ERP system. All messages about sales (from the initial request to closing the deal) are part of the Sales request entity; likewise all messages about purchase (from the initial request to closing the deal) are part of the Purchase request entity.

The ERP system has been configured to calculate KPIs by summing up other values. For example, Cash level is calculated as a total of all transactions that change Cash level – payments for purchases, income from sales, payment of bonuses, initial cash, etc. Turnover and Gross profit is calculated as a total of gross profits and turnovers of specific product types. The values of the most important KPIs in all simulation steps can be exported to an Excel file and analysed later by typical Excel tools like a contingency table or by a data analysis like histograms etc. The negotiation steps can also be exported to Excel in order to analyse the customer and sales representative behavior.

3.2 Simulation Results

We implemented and analysed simulation results from two 90-days scenarios in MAREA to present the functionalities of MAREA. The parameterization of first scenario is depicted in Table 1. We changed some parameters of the simulated company in second scenario and compared the results at the end of the cycle with the first scenario with the intention to acquire more beneficial results. We used tables and graphs with values in both scenarios. After all we decided if our changes were beneficial or not.

We modeled a small-sized trading company for our research. The company deals with retailing of fluorescence colours. It employs 9 sales representatives and 3 purchase representatives. Purchase representatives cooperate with 4 vendors. Vendors supply items in cooperation with purchase representatives to the stock. Our trading company offers 24 items. These 24 items are 14 different colours and most of them are in two variants: 1-liter and 5-liters.

Table 1. First scenario simulation parameters. Source: own.

Parameter	Example
Start date	1.1.2016
Numbers of iterations	90
Global budget	250000
Customers	47
Probability for creating sales request	10
Vendors	4
Probability of disturbance	10
Advertising cost ratio	0.1

We can see the parameters setup of the first scenario in Table 1. First parameter is a starting day. Start date was set on 1.1.2016 in both scenarios. Second parameter is the number of iterations, which in our case means the number of days in both scenarios. We used 90-days, which is exactly 3 months in not leap-year. Each customer has its own budget and after each purchase the budget has to be diminished. Hence, for simulation purpose, there has to be one global budget defined for all customers and this global budget has to be randomized for each customer in the initialization of the customer. The number of customers, who are involved in the process is 47. Probability of creating sales request is a parameter, which affects customers. The larger number is set, the more likely will customers require goods from sales representatives. This is set on 10. The number of vendors is set on 4. Probability of disturbance is another parameter which affects customers. Disturbance decreases quality of service of our sales representatives and at the end negatively affects the customer decision. We set this value on 10. Last parameter is advertising cost ratio. We change marketing costs using this parameter. Increasing of marketing costs more but it can bring increase of sales, which can act positively in final financial balance of the company.

The change we made in the second scenario was in increasing the advertising cost ratio, which was increased to 0.5. We have made these changes according to the experience with the first scenario with the aim to improve company's results and make it more competitive. Some of sales representatives were not effective so we decided to dismiss 3 worst of them. In connection with this dismissing we set the percentage bonus for every sales representative to 10. We also tried to support selling of 5 least selling items from first scenario. This support was done through a reduction of purchase limit prices for these items.

We tracked initial cash level, cash level, turnover, gross profit and profit. Initial cash level is a level of cash at start of the trading cycle. This level was set to 15000. Cash level is calculated as a difference between sales orders and payments. Turnover is a summary of what customers pays for goods. Gross profit is a profit before taxes, depreciation and amortization. And a profit is calculated as a difference between revenues and expenses. These five KPIs were set to 0. Table 2 presents company's results after three months of trading in both scenarios.

Table 2. Key performance indicators for both scenarios. Source: own.

Scenario	Initial cash level	Cash level	Turnover	Gross profit	Profit	Active customers
1	15000	225859.63	1575954.96	946509.23	210859.63	45
2	15000	1221123.41	1797505.09	984155.09	1206123.41	46

Both of our scenarios started at 15000 initial cash level. The cash level increased more than five times in second scenario (Figs. 2 and 3). This change is caused by reducing the number of employees and setting the bonus to 10% of total selling amount for every seller.

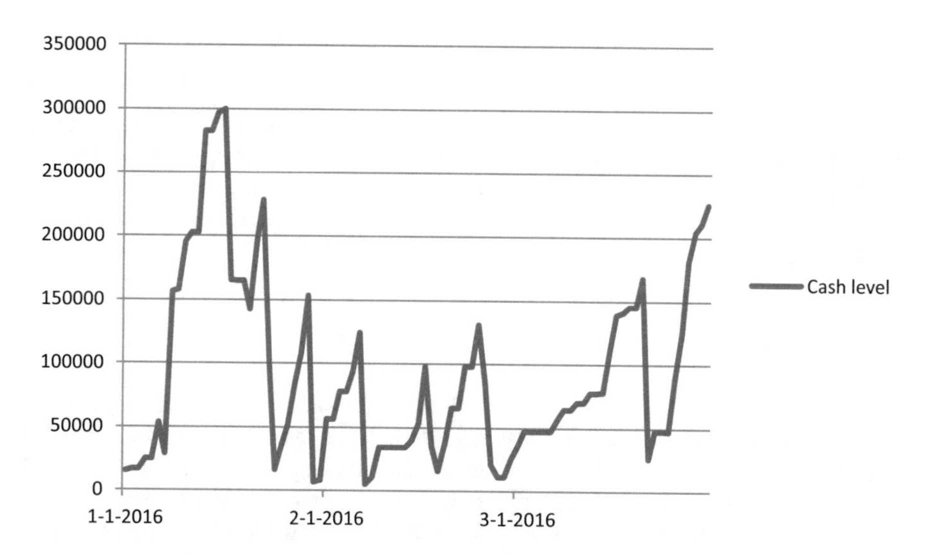

Fig. 2. Cash level in the first scenario. (Source: own)

Cash level is fluctuating and reached value 225859,63 which correspond with the Table 2 as can be seen from a graph of cash level in first scenario (Fig. 2). The reason for the cash level increase more than five times in the second scenario is that the company spent large amount of cash on advertising and also the expanses raised. Bonus payments rose only slightly compared to the sales order.

Cash level in the second scenario (Fig. 3) is more consistent and growing more rapidly than in the first scenario (Fig. 2). Resultant curve reached value of 1221123,41, which confirms the value from Table 2.

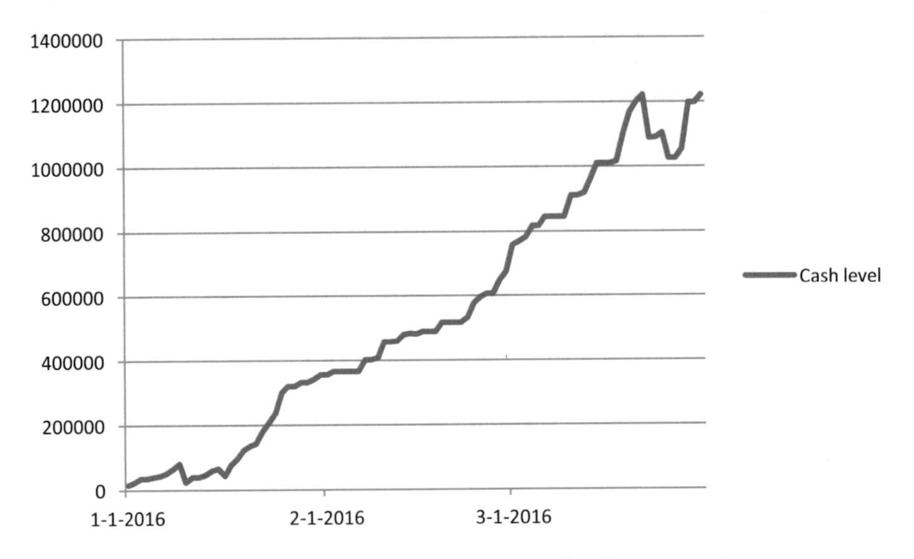

Fig. 3. Cash level in the second scenario. (Source: own)

Obtained results show that simulation results in MAREA are sensitive to initial parameters and simulation results are logically sensitive to changes in parameters. MAREA will be used for implementation of more simulation models of trading companies to be used to support the decision-making processes of companies' management.

4 Conclusion

This paper presents a decision function implementation and simulation experiments in MAREA software framework. Multi-agent system was developed to support simulation experiments dealing with KPIs of a small-sized trading company. The MAREA application serves for the decision support of company's management as well as for educational purposes. It enables users to get familiar with the principles of trading using company model. The setup of the application provides possibilities to edit the company parameters and to run trading simulations. This allows users to analyse trading behavior back-to-back according to the parameters setup. The most important features of MAREA are: configuring options on the lowest level (source code of the agents), availability for distance learning and simulation speed of the framework.

The motivation was to simulate trading processes in order to obtain KPIs (cash level, turnover, profit, etc.) in 90-days of trading behavior. The results obtained show that using such framework can lead to logical outputs. The outputs can be used for improving the decision making process, and to predictive purposes in business companies. Future research will concentrate on more complex setup of real trading company.

Acknowledgement. The work was supported by SGS/19/2016 project of Silesian University in Opava, Czech Republic, Europe called "Advanced mining methods and simulation techniques in business process domain".

References

1. Suchanek, P., Vymetal, D.: Security and disturbances in e-commerce systems. In: Proceedings of the 10th International Conference Liberec Economic Forum 2011 (2011). ISBN: 978-80-7372-755-0
2. Gries, M., Kulkarni, Ch., Sauer, Ch., Keutzer, K.: Comparing Analytical Modeling with Simulation for Network Processors: A Case Study. University of California, Berkeley; Infineon Technologies, Corporate Research, Munich (2011). https://pdfs.semanticscholar.org/83a4/bc3623b74c360512224cb8227bb1dbfd3d51.pdf. Accessed 13 Sep 2016
3. Liu, Y., Trivedi, K.S.: Survivability Quantification: The Analytical Modeling Approach. Department of Electrical and Computer Engineering. Duke University, USA, Durham. http://people.ee.duke.edu/~kst/surv/IoJP.pdf. Accessed 21 Jan 2016
4. Macal, C.M., North, M.J.: Tutorial on agent-based modeling and simulation. In: Proceedings of the Winter Simulation Conference, pp. 2–15 (2005)
5. Scheer, A.-W., Nüttgens, M.: ARIS architecture and reference models for business process management. In: van der Aalst, W., Desel, J., Oberweis, A. (eds.) Business Process Management. LNCS, vol. 1806, pp. 376–389. Springer, Heidelberg (2000). doi:10.1007/3-540-45594-9_24
6. Sierhuis, M.: Modeling and simulating work practice. Ph.D. thesis, University of Amsterdam (2001)
7. Barnett, M.: Modeling & Simulation in Business Process Management. Gensym Corporation, pp. 6–7 (2003). http://w.businessprocesstrends.com/publicationfiles/11-03%20WP%20Mod%20Simulation%20of%20BPM%20-%20Barnett-1.pdf. Accessed 21 Jan 2016
8. Vymetal, D., Sperka, R.: Virtual company simulation for distance learning. In: Proceedings of the Distance Learning Simulation and Communication Conference, Brno, Czech Republic (2013). ISBN: 978-80-7231-919-0
9. Sharma, S., Sharma, J., Devi, A.: Corporate social responsibility: the key role of human resource management. Bus. Intell. J. **2**, 205–213 (2009). http://citeseerx.ist.psu.edu/viewdoc/download?doi=10.1.1.514.7758&rep=rep1&type=pdf. Accessed 13 Sep 2016
10. Vymetal, D., Sperka, R.: MAREA - from an agent simulation application to the social network analysis. Procedia Comput. Sci. **35**, 1416–1425 (2014). doi:10.1016/j.procs.2014.08.198. Proceedings of the Knowledge-Based and Intelligent Information & Engineering Systems 18th Annual Conference – KES 2014, Gdynia, Poland
11. Vymětal, D., Spišák, M., Šperka, R.: An influence of random number generation function to multiagent systems. In: Jezic, G., Kusek, M., Nguyen, N.-T., Howlett, R.J., Jain, L.C. (eds.) KES-AMSTA 2012. LNCS(LNAI), vol. 7327, pp. 340–349. Springer, Heidelberg (2012). doi:10.1007/978-3-642-30947-2_38
12. Spisak, M., Sperka, R.: Financial market simulation based on intelligent agents - case study. J. Appl. Econ. Sci. VI **3**(17), 249–256 (2011). http://www.jaes.reprograph.ro/articles/winter2011/JAES_Fall_2011_online.pdf. Accessed 13 Sep 2016. Romania, Print ISSN: 1843-6110

13. Šperka, R., Spišák, M.: Transaction costs influence on the stability of financial market: agent-based simulation. J. Bus. Econ. Manag. **14**(Suppl. 1), S1–S12 (2013). doi:10.3846/16111699.2012.701227. Taylor & Francis, United Kingdom, London. Print ISSN: 1611-1699
14. Šperka, R., Vymětal, D.: MAREA - an education application for trading company simulation based on REA principles. In: Proceedings of the Information, Communication and Education Application, ICEA 2013. Advances in Education Research, vol. 30, Hong Kong, China, 1–2 November 2013, pp. 140–147. Information Engineering Research Institute (IERI), Delaware (2013). ISBN: 978-1-61275-056-9
15. Foundation for Intelligent Physical Agents, FIPA: FIPA Contract Net Interaction Protocol. In Specification (2002). http://www.fipa.org/specs/fipa00029/SC00029H.html. Accessed 21 Jan 2016
16. Vymětal, D., Scheller, C.: MAREA: multi-agent REA-Based business process simulation framework. In: Proceedings of the International Scientific Conference ICT for Competitiveness, pp. 301–310. OPF SU, Karviná (2012). ISBN: 978-80-7248-731-8

Application of I-Fuzzy Approach to Prediction of Blockability Values in Real-World Data

Elena Mielcová(✉)

Department of Informatics and Mathematics,
School of Business Administration in Karviná, Silesian University in Opava,
Univerzitní náměstí 1934/3, 733 40 Karviná, Czech Republic
mielcova@opf.slu.cz

Abstract. The main aim of this article is to discuss the setting and the construction of blockability values for decision-making agents, whose actions are modeled by transferable utility cooperative game, when possible coalitions of agents are vague – in this case expressed as I-fuzzy coalitions — and test the appropriateness of this approach on real data – in this case the data from voting in the Chamber of Deputies of the Czech Parliament during three subsequent Parliamentary periods. Results of calculations show improvement in values in ex-post I-fuzzy blockability values, as well as in tested predicted I-fuzzy blockability values based on results from preceding periods.

Keywords: I-fuzzy setting · Cooperative game · Blockability value · Parliamentary voting

1 Introduction

Complex multiagent systems are composed of agents, who are considered to make decisions with respect to their designed objectives. These agents are considered to be to some extent autonomous [1]. Agents in complex multiagent system environment can, with respect to their objectives and settings, interact and even cooperate with other agents. The standard descriptions of the multiagent problem of cooperation is given by the theory of cooperative games [2].

In cooperative game theory, agents (referred also as players) of a model game are cooperating in order to increase a mutual profit. In transferable utility games, the profit is expected to be transferred (which means distributed) among players with respect to some coalition agreement. In general, the cooperative game theory studies the mechanism of possible profit distribution, called also solution concepts [3].

The theory of cooperative games was originally derived in order to describe real-world interactions between two or more rational agents. One from many examples of the application of cooperative theory on real-world situations is theory of voting. Rules of every designed election mechanism can be described by

G. Jezic et al. (eds.), *Agent and Multi-Agent Systems: Technology and Applications*, Smart Innovation, Systems and Technologies 74,
DOI 10.1007/978-3-319-59394-4_14

a characteristic function of a transferable utility game. The result of a decision-making mechanism defined as a voting rule is used as a result of voting or election. Complex agent-based models working on the basis of game-theoretical approach can be used to describe real world situations. However, in order to catch all aspect of real life, some sort of vagueness should be incorporated into models. For example, journalists together with statisticians and political sciences try to predict outcome of parliamentary elections. Opinion polls are running till election days, even voters after elections, when leaving polling places, are asked to reveal their cast ballots. Nonetheless, true results significantly differ from estimated ones.

There are several possibilities how to incorporate vagueness into game theory. For example the concept of fuzzy sets [4] is widely is used to describe an uncertainty in coalition creation [5]. In fuzzy sets theory, every fuzzy set is composed of elements characterized by a membership degree to the set; each element is in fact to some extent member of the set and to some extent non-member of the set. However, this concept was not enough in situation, when also undecisive part is present. For example in voting game in ammendment procedure sometimes there are three possible outcomes of voting: 'yes', 'no', and 'abstain'. This situation is fully described using intuitionistic or interval-valued fuzzy sets. The intuitionistic or interval-valued fuzzy set approach is based on the fact that membership, nonmembership as well as an uncertainty part of a set is taken into account [6]. The intuitionistic fuzzy set theory is used to describe game theory concepts in order to more realistically describe real-world concepts [7]. Dubois et al. (2005) [8] discussed terminological difficulties concerning term "an intuitionistic fuzzy set" and proposed use of different term; therefore throughout this text, the term "I-fuzzy" will be used instead of the term "intuitionistic fuzzy".

The main aim of this article is to discuss the setting and the construction of blockability values for decision-making agents, whose actions are modeled by transferable utility cooperative game, when possible coalitions of agents are expressed as I-fuzzy coalitions, and test the appropriateness of this approach on data from voting in the Chamber of Deputies of the Czech Parliament during three subsequent Parliamentary periods.

2 I-Fuzzy Blockability Value for Weighted Voting Game

The idea of blockability value was derived to evaluate influence of players in weighted cooperative games [9]. A cooperative game is in general considered to be a pair (N, v), where $N = \{1, 2 \ldots n\}$ is a set of players, and $v : 2^N \to R$ is a characteristic function of a game, defined on subsets of N with the property $v(\emptyset) = 0$. The function v is called a characteristic function of a game. Any nonempty subset of N is called a coalition. The characteristic function v connects each coalition $K \subset N$ with a real number $v(K) \in R$ representing total profit of coalition K, while $v(\emptyset) = 0$. Any cooperative game is usually denoted as (N, v), or simply only by its characteristic function v.

Any voting of players with different weights can be described as a weighted voting game $[N, w, q]$ for players $N = \{1, 2 \ldots n\}$ with weights

$w = [w_1, w_2, \ldots, w_n]$ and quota q. In voting game a coalition K is winning if its total weight meets or exceeds the quota q:

$$K \text{ wins } \Longleftrightarrow \sum_{i \in K} w_i \geq q \tag{1}$$

The characteristic function of a weighted voting game can obtain only values 0 or 1; $v(K) = 1$ if K wins and $v(K) = 0$ when K loses. From the definition of a weighted voting game it is obvious that the empty coalition never wins, the grand coalition always wins, and any set of winning coalition also wins.

The blockability value was derived in order to easy evaluate a coalitional influence by a real number in order to compare coalitional influence of all players – the higher is the blockability value, the higher is a player's coalition power. The blockability value of the coalition K for a weighted voting game can be is expressed as [9,10]:

$$B_K = \frac{\sum_{T \subset N} v(T) - B^*(K)}{\sum_{T \subset N} v(T)} \tag{2}$$

where $B^*(K) = \sum_{T \subset N} v(T \backslash K)$.

Example: For a game $G = [N = \{A, B, C, D\}, w = (90, 80, 20, 10), q = 101]$ can be easily shown that:
$\sum_{T \subset N} v(T) = v(\emptyset) + v(A) + v(B) + v(C) + v(D) + v(AB) + v(AC) + v(AD) + v(BC) + v(BD) + v(CD) + v(ABC) + v(ABD) + v(ACD) + v(BCD) + v(ABCD) = 0 + 0 + 0 + 0 + 0 + 1 + 1 + 0 + 0 + 0 + 0 + 1 + 1 + 1 + 1 + 1 = 7$, and
$B^*(A) = v(\emptyset) + v(\emptyset)) + v(B) + v(C) + v(D) + v(B) + v(C) + v(D) + v(BC) + v(BD) + v(CD) + v(BC) + v(BD) + v(CD) + v(BCD) + v(BCD) = 0 + 0 + 0 + 0 + 0 + 0 + 0 + 0 + 0 + 0 + 0 + 0 + 0 + 0 + 1 + 1 = 2$.
Similarly, $B^*(B) = 4$, $B^*(B) = 4$, and $B^*(B) = 6$. Then blockability values of all players are:

$$B_A = \frac{\sum_{T \subset N} v(T) - B^*(A)}{\sum_{T \subset N} v(T)} = \frac{7-2}{7} = \frac{5}{7}, B_B = \frac{3}{7}, \; B_C = \frac{3}{7}, \; B_D = \frac{1}{7}.$$

In order to be consistent with possible outcomes of voting, the concept of I-fuzzy coalition was chosen to describe vagueness in Parliamentary voting. I-fuzzy coalition consists of participation as well as non-participation level for each player. Formally, let $N = \{1, 2 \ldots n\}$ be a set of n players. An I-fuzzy coalition $\tilde{C}$ is given by a pair of vectors $\tilde{C} = \langle \mu^C, \nu^C \rangle$ with coordinates $\mu^C = (\mu_1^C, \mu_2^C, \ldots \mu_n^C)$ and $\nu^C = (\nu_1^C, \nu_2^C, \ldots \nu_n^C)$ such that $0 \leq \mu_i^C + \nu_i^C \leq 1$ for all $i \in N$. Then the i-th coordinate of vector μ^C gives a level of membership (participation) of player i in an I-fuzzy coalition $\tilde{C}$; the i-th coordinate of vector ν^C gives a level of nonmembership of player i in an I-fuzzy coalition $\tilde{C}$.

A crisp coalition $S \subset N$ in I-fuzzy notation is represented by I-fuzzy coalition $\tilde{C}^S$ such that $\tilde{C}^S = \langle \mu^S, \nu^S \rangle$ for which $\mu_i^S = 1$ for all $i \in S$ and $\mu_i^S = 0$ for all $i \notin S$ while $\nu_i^S = 1$ for all $i \notin S$ and $\nu_i^S = 0$ for all $i \in S$. Analogically, the empty coalition is expressed as $\tilde{C}^\emptyset = \langle \mu^\emptyset, \nu^\emptyset \rangle$ for which $\mu_i^\emptyset = 0$ and $\nu_i^\emptyset = 1$ for all i, while the grand coalition is of the form $\tilde{C}^N = \langle \mu^N, \nu^N \rangle$ for which $\mu_i^N = 1$ and $\nu^N t_i = 0$ for all i.

Let T be a crisp coalition such that $T \subset N$. Let $\tilde{C}$ be an I-fuzzy coalition. Let e^T be a fuzzy coalition created from $\tilde{C}$ such membership and nonmembership function are at basic levels for players not in coalition T, other membership and nonmembership functions are unchanged. The blockability value of the coalition $K \subset N$ for the referral coalition $\tilde{C}$ can be expressed as:

$$B_K = \frac{\sum_{T \subset N} v(e^T) - B^*(K)}{\sum_{T \subset N} v(e^T)} \tag{3}$$

where $B^*(K) = \sum_{T \subset N} v(e^T \backslash e^K)$. In real-world situation the referral coalition is considered to be a 'typical' coalition for a played game (for example in voting it can be the most probable coalition or an announced coalitional partnership).

In the case of presented real-world data, the characteristic function of the presenter weighted voting game is not known, however actions of all players are revealed. Thus, in order to calculate a characteristic function in the case of the Chamber of Deputies of the Czech Parliament, a probabilistic approach was used, values $v(e^T)$ and $v(e^T \backslash e^K)$ in accord with these three rules:

(1) if the sum of participation levels of party members in coalition T (for simplicity called $\min(T)$) is greater than $(q-1)$, then $v(T) = 1$;
(2) if the sum of participation levels of party members in coalition T plus all values of uncertainty levels of all political parties in T (for simplicity called $\max(T)$) is smaller than q, then $v(T) = 0$;
(3) if none of above cases is true, then the respective value $v(T)$ is calculating as the value of $v(T) = 1 - F(q-1)$, where $F(q-1)$ is the value of a cumulative distribution function for the continuous uniform distribution $(\text{unif}(\min(T), \max(T))$.

3 Data Description

This analysis is based on the data from the Czech Parliament from three subsequent electoral periods: 2002–2006, 2006–2010, 2010–2013. Data for the 2002–2006 parliamentary period covered 13633 voting vectors; data for the 2006–2010 parliamentary period covered 8740 voting vectors; data for the 2010–2013 parliamentary period covered 5895 voting vectors. The outcome of every vote for every member is one possibility from the set {yes, no, abstain, absent}. Every bill to be passed needs at least as many 'yes' votes as quota. Quota is dependent on the number of all present legislators. Therefore, in this analysis the 'abstain' outcomes were reclassified as 'no' outcomes. Basic information on the Czech Parliamentary system as well as the set of all historical votes can be found at the official web site of the Chamber of Deputies of the Czech Parliament URL: www.psp.cz.

During the 2002–2006 parliamentary period there were five political parties operating in the Chamber of Deputies of the Czech Parliament: Civic Democratic Party (ODS), Christian and Democratic Union - Czechoslovak People's

Party (KDU-CSL), Czech Social Democratic Party (CSSD), Communist Party of Bohemia and Moravia (KSCM), and the Freedom Union – Democratic Union (US-DEU). Distribution of seats after 2002 Election in the 2010 Chamber of Deputies of the Czech Republic together with blockability value is given in Table 1.

Table 1. The Chamber of Deputies of the Czech Parliament after 2002 elections. *Source: Official results of 2002 Elections, own calculations.*

Party	CSSD	ODS	KSCM	KDU-CSL	US-DEU
Seats	70	58	41	21	10
Blockability value	0.625	0.375	0.375	0.125	0.125

During the 2006–2010 parliamentary period there were five political parties operating in the Chamber of Deputies of the Czech Parliament: Civic Democratic Party (ODS), Christian and Democratic Union - Czechoslovak People's Party (KDU-CSL), Green Party (SZ), Czech Social Democratic Party (CSSD), and Communist Party of Bohemia and Moravia (KSCM). Distribution of seats in the 2006–2010 Chamber of Deputies of the Czech Republic together with blockability value is given in Table 2.

Table 2. Distribution of seats and calculated blockability value in the Chamber of Deputies of the Czech Parliament after 2006 elections. *Source: Official results of 2006 Elections, own calculations.*

Party	ODS	CSSD	KSCM	KDU-CSL	SZ
Seats	81	74	26	13	6
Blockability value	0.6	0.467	0.467	0.067	0.067

During the 2010–2013 parliamentary period there were five political parties operating in the Chamber of Deputies of the Czech Parliament: Civic Democratic Party (ODS), Czech Social Democratic Party (CSSD), Communist Party of Bohemia and Moravia (KSCM), TOP09 and Veci Verejne (VV) – respective distribution of seats and the blockability value is given in Table 3.

4 Ex-post Blockability Value

In order to evaluate appropriateness of calculated results, the coefficient of political party success was calculated. The political party success coefficient is defined as the ratio of decisions in the Chamber of Deputies that were the same as the party decisions to all decisions during the parliamentary period. The party decision is derived from the votes of party members using simple majority rule.

Table 3. Distribution of seats and calculated blockability value in the Chamber of Deputies of the Czech Parliament after 2010 elections. *Source: Official results of 2010 Elections, own calculations.*

Party	CSSD	ODS	TOP09	KSCM	VV
Seats	56	53	41	26	24
Blockability value	0.5	0.5	0.25	0.25	0.25

The coefficient of party success can reach the values from the interval $[0, 1]$; the higher the coefficient, the higher ratio of party decisions was the same as the whole voting body decision. The coefficient of voting success is influenced by the vote of other members of parliament. Calculated party success indices for all three parliamentary periods are given in Table 4.

Table 4. Coefficient of success in the Chamber of Deputies of the Czech Parliament during three parliamentary periods. *Source: Own calculations.*

2002–2006					
Party	CSSD	ODS	KSCM	KDU-CSL	US-DEU
Party success	0.930	0.573	0.753	0.852	0.822
2006–2010					
Party	ODS	CSSD	KSCM	KDU-CSL	SZ
Party success	0.808	0.812	0.721	0.749	0.678
2010–2013					
Party	CSSD	ODS	TOP09	KSCM	VV
Party success	0.592	0.943	0.952	0.567	0.831

The easiest way how to evaluate the reliability of the party blockability value with reality is to calculate the correlation coefficient between the party success and the blockability value. The calculating coefficient values are 0.094, 0.690, and −0.046 for the 2002–2006, 2006–2010, and 2010–2013 parliamentary periods, respectively. None of the correlation coefficient is statistically significant. Hence, the adjustment to blockability value has to be done in order to adjust to a real-world situation. One of the possibilities how to adjust calculated blockability value is to estimate blockability value with respect to I-fuzzy setting.

As this text deal with possible prediction of future blockability value when only election results and previous votes are known, it should be important to discuss ex-post blockability value under I-fuzzy setting in order to show that this approach could bring reliable results. In order to find the membership and nonmembership function of a 'referral coalition', the party cohesion concept was used. The membership function of a political party is calculated as the average participation of that political party in winning coalition; similarly, the nonmembership function of a political party is calculated as the average participation of that political party in losing coalitions. Respective membership and nonmembership values are given in Table 5.

Table 5. Membership and nonmembership values for political parties in the Chamber of Deputies of the Czech Parliament during three parliamentary periods. *Source: Own calculations.*

2002–2006					
Party	CSSD	ODS	KSCM	KDU-CSL	US-DEU
Membership value	0.731	0.419	0.628	0.683	0.545
Nonembership value	0.103	0.364	0.248	0.155	0.140
2006–2010					
Party	ODS	CSSD	KSCM	KDU-CSL	SZ
Membership value	0.604	0.602	0.605	0.535	0.523
Nonembership value	0.196	0.205	0.278	0.196	0.217
2010–2013					
Party	CSSD	ODS	TOP09	KSCM	VV
Membership value	0.392	0.672	0.761	0.453	0.592
Nonembership value	0.354	0.087	0.072	0.407	0.156

Calculated blockability values of political parties based on calculated membership and nonmembership values of 'referral coalition' are given in Table 6.

Correlation coefficient between the party successes and the blockability values are equal to 0.17, 0.91 and 0.68, for the 2002–2006, 2006–2010, and 2010–2013 periods, respectively. The second correlation coefficient is statistically significant at 5% level of significance, other ones are not significant, however show improvement comparing to original blockability values. The problem with result is, that it does not take into account created coalitions (both governmental and opposition ones).

5 Blockability Value Predictions

There are several possibilities how to estimate membership and nonmemberhip value of a political party in the 'referral coalition'. The easiest way is to set all membership and nonmembership values equal to averages of these values in preceding period, as given in Table 7 (rows Membership Value 1, and Nonmembership Value 1). Another possibility is to take all membership and nonmembership values from the previous period for political parties present in both parliaments, and average values for political parties not present in previous period parliament (rows Membership Value 2, and Nonmembership Value 2 in

Table 6. Ex-ante I-fuzzy Blockability values for political parties in the Chamber of Deputies of the Czech Parliament during three parliamentary periods. *Source: Own calculations.*

2002–2006					
Party	CSSD	ODS	KSCM	KDU-CSL	US-DEU
I-fuzzy blockability value	1	0.580	0.514	0.339	0.126
2006–2010					
Party	ODS	CSSD	KSCM	KDU-CSL	SZ
I-fuzzy blockability value	1	1	0.042	0.042	0.014
2010–2013					
Party	CSSD	ODS	TOP09	KSCM	VV
I-fuzzy blockability value	0.621	0.889	0.759	0.282	0.395

Table 7). Calculated blockability values based on electoral results and estimated membership and nonmembership values for two presented scenarios are given in Table 7.

Table 7. Distribution of seats and calculated predicted blockability values in the Chamber of Deputies of the Czech Parliament after 2006 elections. *Source: Official results of 2006 Elections, own calculations.*

Party 2006–2010	ODS	CSSD	KSCM	KDU-CSL	SZ
Seats	81	74	26	13	6
Membership value 1	0.601	0.601	0.601	0.601	0.601
Nonembership value 1	0.202	0.202	0.202	0.202	0.202
Blockability value 1	1	0.998	0.044	0.044	0.017
Membership value	0.731	0.419	0.628	0.683	0.601
Nonembership value	0.103	0.364	0.248	0.155	0.202
Blockability value 2	0.975	1	0.118	0.091	0.044

Correlation coefficients of predicted values with original blockability value, ex-post I-fuzzy blockability value, and coefficient of success are given in Table 8. Correlations of predicted values with success coefficient, as well as with ex-ante blockability value are close to 1, these results are statistically significant at 5% level of significance. The result is caused by favorably distributed seats after 2006 elections.

In order to estimate blockability value for 2010–2013 parliamentary period, the membership and nonmemberhip value of a political parties in the 'referral coalition' were estimated with respect to two previous periods. Calculated blockability values based on electoral results and estimated membership and nonmembership values for two presented scenarios are given in Table 9.

Table 8. Pairwise correlation coefficients between calculated values for 2006–2010 Parliamentary period. All correlation coefficients higher than 0.879 are statistically significant at 5% level of significance. *Source: Own calculations.*

	Success	Blockability	IF blockability	Prediction 1	Prediction 2
Success	1.0000	0.6870	0.9058	0.9055	0.9132
Blockability	0.6870	1.0000	0.7387	0.7387	0.7546
IF blockability	0.9058	0.7387	1.0000	1.0000	0.9993
Prediction 1	0.9055	0.7387	1.0000	1.0000	0.9992
Prediction 2	0.9132	0.7546	0.9993	0.9992	1.0000

Table 9. Distribution of seats and calculated predicted blockability values in the Chamber of Deputies of the Czech Parliament after 2010 elections. *Source: Official results of 2010 Elections, own calculations.*

Party 2010–2013	CSSD	ODS	TOP09	KSCM	VV
Seats	56	53	41	26	24
Membership value 1	0.587	0.587	0.587	0.587	0.587
Nonembership value 1	0.210	0.210	0.210	0.210	0.210
Blockability value 1	0.839	0.815	0.581	0.385	0.367
Membership value	0.666	0.511	0.587	0.616	0.587
Nonembership value	0.154	0.280	0.210	0.263	0.210
Blockability value 2	0.890	0.765	0.592	0.384	0.377

Table 10. Pairwise correlation coefficients between calculated values for 2006–2010 Parliamentary period. All correlation coefficients higher than 0.879 are statistically significant at 5% level of significance. *Source: Own calculations.*

	Success	Blockability	IF blockability	Prediction 1	Prediction 2
Success	1.0000	−0.0465	0.6811	0.1219	0.0335
Blockability	−0.0465	1.0000	0.6032	0.9279	0.9048
IF Blockability	0.6811	0.6032	1.0000	0.7940	0.7304
Prediction 1	0.1219	0.9279	0.7940	1.0000	0.9873
Prediction 2	0.0335	0.9048	0.7304	0.9873	1.0000

Correlation coefficients of predicted values with original blockability value, ex-post I-fuzzy blockability value, and coefficient of success are given in Table 10. Results are better off comparing to real values than crisp blockability values; however, correlations of predicted values with success coefficient, as well as with ex-ante blockability value are not statistically significant. The main reason of the failure is the created coalition of three political parties – ODS, TOP09, and VV. The similarity in voting of these three players increased their measured success coefficient disproportional to their weights. This increase was not covered in predicted membership and nonmembership values.

6 Concluding Remarks

The main aim of this article was to apply I-fuzzy blocking values on real voting data, and use the concept of blocking values to predict future outcomes. Real voting data were represented by a voting data from the Chamber of Deputies of the Parliament of the Czech Republic for three consequent periods: 2002–2006, 2006–2010 and 2010–2013 periods. Results of the ex-ante and ex-post blocking values are compared with overall voting outcome represented by coefficient of success. For the three referral periods 2002–2006, 2006–2010, and 2010–2013, the correlation coefficients between ex-ante blockability values and party voting success were 0.094, 0.690, and −0.046, respectively. None of them was statistically significant at 5% level of significance. The same correlation coefficient for ex-post I-fuzzy blockability value were equal to 0.17, 0.91 and 0.68; the second correlation coefficient was statistically significant. Predicted values also show an improvement in results when uncertainty issues are considered, however the expected coalition formation (possible future coalition treaty) should be taken into account.

Acknowledgments. This paper was supported by the Ministry of Education, Youth and Sports Czech Republic within the Institutional Support for Long-term Development of a Research Organization in 2017.

References

1. Wooldridge, M.J.: An Introduction to Multiagent Systems. Wiley, New York (2002)
2. Shoham, Y., Leyton-Brown, K.: Multiagent Systems: Algorithmic, Game-Theoretic, and Logical Foundations. Cambridge University Press, Cambridge (2009)
3. Myerson, R.B.: Game Theory: Analysis of Conflict. Harvard University Press, Cambridge (1991)
4. Zadeh, L.A.: Fuzzy sets. Inf. Control **8**(3), 338–353 (1965)
5. Aubin, J.P.: Cooperative fuzzy games. Math. Oper. Res. **6**(1), 1–13 (1981)
6. Atanassov, K.T.: Intuitionistic fuzzy sets. Fuzzy Sets Syst. **20**(1), 87–96 (1986)
7. Li, D.-F.: Decision and Game Theory in Management with Intuitionistic Fuzzy Sets. SFSC, vol. 308. Springer, Heidelberg (2014)
8. Dubois, D., Gottwald, S., Hajek, P., Kacprzyk, J., Prade, H.: Terminological difficulties in fuzzy set theory - the case of "Intuitionistic Fuzzy Sets". Fuzzy Sets and Syst. **156**(3), 485–491 (2005)
9. Kojima, K., Inohara, T.: Methods for comparison of coalition influence on games in characteristic function form. Appl. Math. Comput. **217**, 4047–4050 (2010)
10. Kojima, K., Inohara, T.: Coalition values derived from methods for comparison of coalition influence for games in characteristic function form. Appl. Math. Comput. **219**, 1345–1353 (2012)

Generalized Dynamic Model of Rating Alternatives by Agents with Interactions

Radomír Perzina(✉) and Jaroslav Ramík

Faculty of Business Administration, Silesian University in Opava, Karviná, Czech Republic
{perzina,ramik}@opf.slu.cz

Abstract. Business process simulation models usually incorporate several essential components that reflect customer behavior for modeling system inputs and outputs and ranking and/or rating given alternatives. In this paper we deal with a general dynamic system of rating a number of alternatives given by pairwise comparison matrices on an alo-group. This system is based on a parametrized agent-based simulation with interactions among agents which is able to replicate various types of processes, e.g. financial market evaluation, evaluation of products' demand and supply, evaluation of political parties in general elections, evaluation of universities etc. A simple simulation experiment is presented and discussed.

Keywords: Dynamic system · Agents · Ranking alternatives · Pairwise comparison matrix on alo-group · Simulation

1 Introduction

An agent-based model is a collection of autonomous decision-making entities - agents. Each agent individually assesses its situation and makes decisions on the basis of some rules. Agents may execute various behaviors appropriate for the system they represent, for example, production system, consuming, or selling system, and many others. A significant feature of agent-based modeling relies on the power of computers to explore dynamics out of the reach of pure mathematical methods by means of repetitive competitive interactions between the agents, see [1–4]. In addition, agents may be capable of evolving, allowing unanticipated behaviors to emerge, see [3,5].

A process simulation models usually incorporate several basic components that reflect agent behavior for modeling system inputs and outputs and ranking and/or rating given alternatives, see e.g. [3,6–8]. In this paper we deal with a dynamic system of rating given number of alternatives based on alo-groups where agent-based simulations with interactions among agents are dependent on some parameters. The system of rating alternatives is able to replicate various processes, e.g. financial market modeling, see [7,8], auction models [2], demand

G. Jezic et al. (eds.), *Agent and Multi-Agent Systems: Technology and Applications*, Smart Innovation, Systems and Technologies 74,
DOI 10.1007/978-3-319-59394-4_15

and supply models [6], evaluation of political parties in general elections, evaluation of universities or other public institutions, etc.

The structure of the paper can briefly be described as follows. In Sect. 2 the basic problem is formulated as well as necessary notions and terminology concerning alo-groups. In each time moment, a finite set of alternatives is ranked by each agent from the finite set of agents. Then, the total ranking is calculated. In the course of time, the individual agents interact with each other according to the given system of rules. A detailed agent-based model is described in Sect. 3. Finally, a simple simulation experiment is presented and discussed in Sect. 4. The well known additive alo-group is applied. The conclusion section finalizes the paper.

2 Preliminaries

We shall deal with the following problem: Let $X = \{x_1, x_2, ..., x_n\}$ be a finite set of alternatives ($n > 1$). We consider a set of agents $K = \{1, 2, ..., |K|\}$, e.g. brokers, customers, electors, students etc. Here, by symbol $|K|$ we denote the (finite) number of elements of the set K. The whole group of agents K is composed of various sub-groups of similar agents. In every time moment the agents may interact (communicate) among each others and switch from one sub-group to some other sub-group.

Each agent $k \in K$ makes successively his/her decisions in every discrete time moment $t = 1, 2, ..., T$ by a pairwise comparisons matrix (PC matrix) $A(k,t)$. The aim is to get *global rating*, or, *group rating*, of the given alternatives in each time moment t, using the information given by each agent k in the form of an $n \times n$ *individual pairwise comparisons matrix*

$$A(k,t) = \{a_{ij}(k,t)\}. \tag{1}$$

In [9], given time t each agent k evaluates the pair of alternatives x_i, x_j by a positive real number $a_{ij}(k,t)$, for all i and j. In this paper, we extend this approach to assume that the element $a_{ij}(k,t)$ of the pairwise comparisons matrix $A(k,t)$ belongs to a more general structure then positive real numbers with the operation of multiplication. Particularly, we shall assume that the elements $a_{ij}(k,t)$ belong to an Abelian linearly ordered group G, (shortly *alo-group*) $G \subset \mathbf{R}$, i.e. a subset of the set of real numbers. This approach enables us to apply an agent based approach not only to multiplicative system as described in [3], but also to apply additive systems, fuzzy systems or other systems. Bellow, we shortly remind some necessary terminology associated with alo-groups.

An *abelian group* is a set, G, together with an operation $\odot$ (read: operation *odot*) that combines any two elements $a, b \in G$ to form another element in G denoted by $a \odot b$, see [10]. The symbol $\odot$ is a general placeholder for a concretely given operation. $(G, \odot)$ satisfies the following requirements known as the *abelian group axioms*, particularly: *commutativity*, *associativity*, there exists an *identity element* $e \in G$ and for each element $a \in G$ there exists an element $a^{(-1)} \in G$ called the *inverse element to* a.

The *inverse operation* $\div$ to $\odot$ is defined for all $a, b \in G$ as follows

$$a \div b = a \odot b^{(-1)}. \tag{2}$$

An ordered triple $(G, \odot, \leq)$ is said to be *abelian linearly ordered group, alo-group* for short, if $(G, \odot)$ is a group, $\leq$ is a linear order on G, and for all $a, b, c \in G$

$$a \leq b \text{ implies } a \odot c \leq b \odot c. \tag{3}$$

If $\mathcal{G} = (G, \odot, \leq)$ is an alo-group, then $G \subset \mathbf{R}$ is naturally equipped with the order topology induced by $\leq$ and $G \times G$ is equipped with the related product topology. We say that $\mathcal{G}$ is a *continuous alo-group* if $\odot$ is continuous on $G \times G$.

$\mathcal{G} = (G, \odot, \leq)$ is *divisible* if for each positive integer n and each $a \in G$ there exists the (n)-th root of a denoted by $a^{(1/n)}$, i.e. $\left(a^{(1/n)}\right)^{(n)} = a$.

Because of the associative property, the operation $\odot$ can be extended by induction to n-ary operation: $a^{(n)} = a \odot a \odot ... \odot a$. By divisibility, (n)-th power of a can be extended to (c)-th power of a, $a^{(c)}$, where $c \in \mathbf{R}$. It is well known fact, that a continuous alo-group of real numbers is divisible and that it must be an open interval.

We define a "multiplication" operation "$\bullet$" as follows: For all $a \in G, c \in \mathbf{R}:$

$$c \bullet a = a^{(c)}. \tag{4}$$

Then $\mathcal{H} = (G, \odot, \bullet, \leq)$ is the *Riesz space*, with the following properties, see [11]:

- $(G, \odot, \bullet)$ is a vector space over the field $\mathbf{R}$;
- $(G, \leq)$ is a lattice;
- for every $a, b \in G$, $c \geq 0, a \leq b$ implies $c \bullet a \leq c \bullet b$.

Let $\mathcal{G} = (G, \odot, \leq)$ be an alo-group. As every alo-group $\mathcal{G}$ is a lattice ordered group, there exists $\max\{a, b\}$, for each pair $(a, b) \in G \times G$. Nevertheless, a nontrivial alo-group $\mathcal{G} = (G, \odot, \leq)$ has neither the greatest element nor the least element. Then, function $\|.\| : G \to G$ defined for each $a \in G$ by

$$\|a\| = \max\{a, a^{(-1)}\} \tag{5}$$

is called a $\mathcal{G}$-*norm.*

The operation $d : G \times G \to G$ defined by $d(a, b) = \|a \div b\|$ for all $a, b \in G$ is called a $\mathcal{G}$-*distance*. Next, we present the well known examples of alo-groups of real numbers $\mathbf{R}$, see also [10] or [12].

Example 1: Additive alo-group $\mathcal{R} = (\mathbf{R}, +, \leq)$ is a continuous alo-group with: $e = 0,\ a^{(-1)} = -a,\ a^{(n)} = a + a + ... + a = n.a.$

Example 2: Multiplicative alo-group $\mathcal{R}^+ = (]0, +\infty[, \bullet, \leq)$ is a continuous alo-group with: $e = 1,\ a^{(-1)} = a^{-1} = 1/a,\ a^{(n)} = a^n$. Here, by $\bullet$ we denote the usual operation of multiplication.

Example 3: Fuzzy additive alo-group $\mathcal{R}_a = (\mathbf{R}, +_f, \leq)$, see [12], is a continuous alo-group with: $a +_f b = a + b - 0.5$, $e = 0.5, a^{(-1)} = 1 - a, a^{(n)} = n.a - (n-1)/2$.

Example 4: Fuzzy multiplication alo-group $]\mathbf{0}, \mathbf{1}[_\mathbf{m} = (]0, 1[, \bullet_f, \leq)$, see [10], is a continuous alo-group with:$a \bullet_f b = \frac{ab}{ab + (1-a)(1-b)}, e = 0.5, a^{(-1)} = 1 - a.$

3 Agent-Based Model

Now, let us return back to our original rating problem. If $a_{ij}(k,t) > e$, then x_i "is better than" x_j. Here, e is the identity element of a real continuous alo-group $\mathcal{G}$. The higher is $a_{ij}(k,t)$, the stronger is agent's evaluation that x_i "is better than" x_j.

On the other hand, if $a_{ij}(k,t) < e$, then x_j "is better then" x_i. The lower is the value of $a_{ij}(k,t)$, the stronger the evaluation that x_j "is better than" x_i.

If $a_{ij}(k,t) = e$, then both alternatives x_j, x_i are evaluated equally.

PC matrix (1) is assumed to be *reciprocal*, which is a natural requirement, see e.g. [10,12,13]. Hence, we have

$$a_{ji}(k,t) = a_{ij}(k,t)^{(-1)}, \text{ for all } i,j \in \{1,2,...,n\}, k \in K, t = 1,2,...,T. \tag{6}$$

The global rating of the alternatives $x_1, x_2, ..., x_n$ in time t is associated with the *global priority vector* $w(t) = (w_1(t), ..., w_n(t))$ which is calculated from the *global PC matrix* $A(t) = \{a_{ij}(t)\}$, by aggregation (i.e. by the $\odot$-average over all agents) of individual PC matrices as follows, see [12]:

$$a_{ij}(t) = \left(\bigodot_{k=1}^{|K|} a_{ij}(k,t) \right)^{\left(\frac{1}{|K|}\right)}, \text{ for all } i,j \in \{1,2,...,n\}, t = 1,2,...,T. \tag{7}$$

The weights w_j of the global priority vector $w(t) = (w_1(t), ..., w_n(t))$ are calculated as row geometric averages of the global PC matrix as follows [12]:

$$w_i(t) = \kappa(t) \left(\bigodot_{j=1}^{n} a_{ij}(t) \right)^{\left(\frac{1}{n}\right)}, \text{ for all } i \in \{1,2,...,n\}, t = 1,2,...,T. \tag{8}$$

Here, $\kappa(t)$ is a normalizing factor as the global priority vector should be normalized in every time moment $t = 1,2,...,T$, i.e.

$$\bigodot_{j=1}^{n} w_j(t) = e, \text{ for } t = 1,2,...,T. \tag{9}$$

From (8) and (9) we obtain easily

$$\kappa(t) = \left(\bigodot_{j=1}^{n} \left(\bigodot_{i=1}^{n} a_{ij}(t) \right)^{\left(\frac{1}{n}\right)} \right)^{(-1)}, \text{for all } t = 1,2,...,T. \tag{10}$$

The global priority vector is associated with the ranking of alternatives as follows:

$$\text{If } w_i(t) > w_j(t) \text{ then } x_i \succ x_j,$$

where $\succ$ stands for "is better than" [12,13].

The rating of alternatives is given directly by the value of elements $w_i(t)$ of global priority vector $w(t)$. Our final task is to analyze the time series of global ratings of individual alternatives and assess some regularities and/or irregularities in their behavior.

In the course of time, the individual agents interact with each other, see [9], personally, by social networks or otherwise. Particularly, in time $t = 1, 2, ..., T$, each agent $k \in K$ belongs to exactly one of s agent-types, $s \in S = \{1, 2, ..., |S|\}$, where $|S| \geq 1$, i.e. he/she belongs to exactly one of disjoint sets $K_1(t), K_2(t), ...,$ $K_{|S|}(t)$ satisfying $K_1(t) \cup K_2(t) \cup ... \cup K_{|S|}(t) = K$, $K_r(t) \cap K_s(t) = \emptyset, r \neq s$. The set of agents $K = \{1, 2, ..., |K|\}$ is supposed to be constant over time $t = 1, 2, ..., T$.

Each agent $k \in K$ is characterized by the equation of dynamics of the changes of evaluations of the alternatives as follows:

$$a_{ij}(k, t+1)) = a_{ij}(k, t) \odot [a_{ij}(k, t) \div a_{ij}(k, t-1)]^{(c)} \odot m(k, t), \qquad (11)$$

$$a_{ji}(k, t+1)) = a_{ij}(k, t+1)^{(-1)}, \text{for all } i, j \in \{1, 2, ..., n\}, i < j, \text{ and } k \in K, \quad (12)$$

for all $t = 1, 2, ..., T$. Here, (12) is the reciprocity condition, moreover, for $i \in \{1, 2, ..., n\}$ we have $a_{ii}(k, t+1)) = e$, $c \in \mathbf{R}$ is a parameter, $m(k, t)$ is a *random* member (or, *error*) with a given probability distribution, e.g. the normal one $N(e, \sigma(k, t))$.

From (11) it follows that the pairwise comparisons of alternatives given by individual agents in time $t+1$ depend on the previous evaluations in time t and, on a relative increment of evaluations between time t and $t-1$, moreover, a small random value (dependent on a parameter σ) is added to the right hand side of equation (11) in order to express small non-specific effects around identity element e.

In each time $t = 1, 2, ..., T$, the agents may meet, or interact, at random, and there is a probability that one agent may convince the other agent to follow his/her opinion. In addition, there is also a small probability that an agent changes his/her opinion independently. A key property of this model is that direct interactions between heterogeneous agents may lead to substantial opinion swings from the set of type $K_r(t)$, into a set of type $K_s(t)$, or, vice versa. Here, $r, s \in S, r \neq s$. As an example, consider $|S| = 2$, $K_1(t)$ is the set of optimistic traders, $K_2(t)$ is the set of pessimistic traders.

The above mentioned swings from the set of type $K_r(t)$, into a set of type $K_s(t)$, are modeled as follows.

For each sub-group of agents $K_s(t)$ of type $s \in S, t = 1, 2, ..., T$, we calculate the *group PC matrix* $A^s(t) = \{a^s_{ij}(t)\}$ as $\odot$-mean as follows:

$$a^s_{ij}(t) = \left(\bigodot_{k \in K_s(t)} a_{ij}(k, t) \right)^{\left(\frac{1}{|K_s(t)|}\right)}, \text{ for all } i, j \in \{1, 2, ..., n\}. \qquad (13)$$

Here again, by $|K_s(t)|$ we denote the number of elements (i.e. agents) in the set $K_s(t)$.

Given $t \in \{1, 2, ..., T\}$, $s \in S$. For each $k \in K$ we compute the "distance" $d(k, s, t)$ of PC matrix $A(k, t)$ from $A^s(t)$, i.e.

$$d(k, s, t) = \|\{a_{ij}(k, t) \div a_{ij}^s(t)\}\|. \tag{14}$$

Here, $\|...\|$ is a matrix norm, e.g. for an $n \times n$ matrix $A = \{a_{ij}\}$, we use the norm, see e.g. [12]:

$$\|A\| = \max\{\max\{a_{ij}, a_{ij}^{(-1)}\}|i, j = 1, ..., n\}. \tag{15}$$

For $k \in K_s(t)$, denote

$$d(k, s^*, t) = \min\{\ d(k, s, t)|s \in \{1, 2, ..., S\}\}. \tag{16}$$

If $d(k, s, t) > d(k, s^*, t)$, then in time $t + 1$ the agent k switches from type s to type s^*, i.e.

$$K_s(t + 1) = K_s(t) - \{k\}, K_{s^*}(t + 1) = K_{s^*}(t) \cup \{k\}. \tag{17}$$

Therefore, the number of agents will change as follows:

$$|K_s(t + 1)| = |K_s(t)| - 1, \text{ and } |K_{s^*}(t + 1)| = |K_{s^*}(t)| + 1. \tag{18}$$

If $d(k, s, t) \leq d(k, s^*, t)$, then, in time $t+1$, the agent k does not switch into any other type group. In this way we perform all switches of agents in K, i.e. we obtain new sets $K_s(t + 1), s \in \{1, 2, ..., S\}$ in time moment $t + 1$.

The algorithm continues in time moments $t+1, t+2, ..., T$ by computing new group PC matrices, new global PC matrix, global rating of alternatives, etc.

4 Simulation Experiment

Our problem formulated in Sects. 2 and 3 will be illustrated on a simulation experiment. Let $X = \{x_1, x_2, ..., x_n\}$ be a set of n alternatives. Here, we assume the additive alo-group $\mathcal{R} = (\mathbf{R}, +, \leq)$ from Example 1, i.e. $\odot = +$. Then, formulas (7)–(11) can be then reformulated as follows:

$$a_{ij}(t) = \frac{1}{|K|}\sum_{k=1}^{|K|} a_{ij}(k, t), \text{ for all } i, j \in \{1, 2, ..., n\}, t = 1, 2, ..., T; \tag{19}$$

$$w_i(t) = \kappa(t) + \frac{1}{n}\sum_{j=1}^{n} a_{ij}(t), \text{ for all } i \in \{1, 2, ..., n\}, t = 1, 2, ..., T; \tag{20}$$

$$\sum_{j=1}^{n} w_j(t) = 0, \text{ for all } t = 1, 2, ..., T; \tag{21}$$

$$\kappa(t) = -\frac{1}{n}\sum_{j=1}^{n}\sum_{i=1}^{n} a_{ij}(t), \text{ for all } t = 1, 2, ..., T; \tag{22}$$

Dynamic changes of the individual PC matrix (11) are reformulated by the following equations:

$$a_{ij}(k,t+1)) = a_{ij}(k,t) + c.[a_{ij}(k,t) - a_{ij}(k,t-1)] + m(k,t), \tag{23}$$

$$a_{ji}(k,t+1)) = -a_{ij}(k,t+1), \text{ for all } i,j \in \{1,2,...,n\}, i < j, \text{ and } k \in K,$$

where $t = 1, 2, ..., T$. Moreover, for $i \in \{1, 2, ..., n\}$ we set the diagonal elements of the PC matrix as identity elements, i.e. $a_{ii}(k, t + 1)) = 0$. Here, $m(k, t)$ is a random member (error) with normal distribution $N(0, \sigma(k, t))$, where $\sigma(k, t) = \sigma$, and c is a parameter.

For a concrete example, we consider $n = 4$, the set of agents $K = \{1, 2, ..., 15\}$, and $T = 100$. Each agent $k \in K$ makes successively his/her decisions in every discrete time moment $t = 1, 2, ..., 100$ by a PC matrix $A(k, t)$ with the dynamic equation (23). The aim is to investigate global rating of the alternatives in each time moment t, particularly in the final time $t = 100$, which can be interpreted as a time of prediction of ranking of alternatives. To this aim we use information given by each agent k in the form of a 4×4 individual PC matrix.

The simulation starts with 4×4 PC matrix for $k \in K$:

$$A(k,0) = \{a_{ij}(k,0)\} = \begin{pmatrix} 0 & 2 & 4 & 6 \\ -2 & 0 & 2 & 4 \\ -4 & -2 & 0 & 2 \\ -6 & -4 & -2 & 0 \end{pmatrix} = A(k,1) = \{a_{ij}(k,1)\}. \tag{24}$$

Here, we use the well known additive scale $-9, -8, ..., -1, 0, 1, 2, ..., 8, 9$, see [14], with the following interpretation of comparisons of pairs x versus y:
0 ... x is equally important to y,
3 ... x is slightly more important to y,
5 ... x is strongly more important to y,
7 ... x is very strongly more important to y,
9 ... x is absolutely more important to y.
Values 1, 2, 4, 6 and 8 are intermediate values, e.g. 4 means that x is between slightly more important and strongly more important to y. The negative values denote the reciprocal importances. The priority vector $w(1) = (3; 1; -1; -3)$, corresponding to $A(k, 0)$, generates the rating of alternatives: $x_1 > x_2 > x_3 > x_4$.

In time t, each agent belongs to exactly one of $|S| = 2$ agent-types, i.e. two disjoint sub-groups $K_1(t), K_2(t)$ satisfying $K_1(t) \cup K_2(t) = K$, $K_1(t) \cap K_2(t) = \emptyset$. The set of agents $K = \{1, 2, ..., 15\}$, is constant in time $t = 1, 2, ..., 100$. The initial matrices $A(k, 0), A(k, 1)$, $k = 1, 2, 3, ..., 15$, are given by (24). The random error m has the normal distribution $N(0, \sigma)$. In the time moment $t = 1$, the agents belong to 2 initial types.

Again, the dynamics of the changes of evaluations of the alternatives by agent $k \in K_s(t)$ is given by formula (23) with the values of $\sigma(k, t) = \sigma$ for all k and t, e.g. $\sigma = 1.0$, or, $\sigma = 3.0$. Moreover, c is a parameter with the value less than 1, e.g. $c = 0.1$, or, $c = 0.3$. For computing distances of an agent's PC matrix to the global PC matrix, we use norm (15). Changes of number of individual agents of a given type are given by formulas (17) and (18). For an illustration, the results of simulation computations for various values of parameter σ and c in the form of time series of the global weights of the alternatives are depicted.

In Fig. 1, we can see that the global weights $w_i(t), i = 1, 2, 3, 4$, of alternatives x_i remain approximately constant in the course of time t, $t = 1, 2, ..., 100$. Here, we consider a small error m, i.e. small values of the parameters $\sigma = 1.0, c = 0.1$ Consequently, in the final time moment $t = 100$ the vector of weights will not change much: $w(100) = (3.1; 1.2; -0 : 5; -3.8)$, with the corresponding ranking of alternatives: $x_1 > x_2 > x_3 > x_4$. In Fig. 2, the number of agents of both types are depicted depending on time t.

Moreover, in Figs. 3 and 4, we have increased the value of the error to the value $\sigma = 3.0$, and also $c = 0.3$. As we can see, even though in the final time moment $t = 100$ the vector of weights is as follows: $w(100) = (5.8; 4.2; 3.2; -13.2)$, with the same corresponding ranking of alternatives:

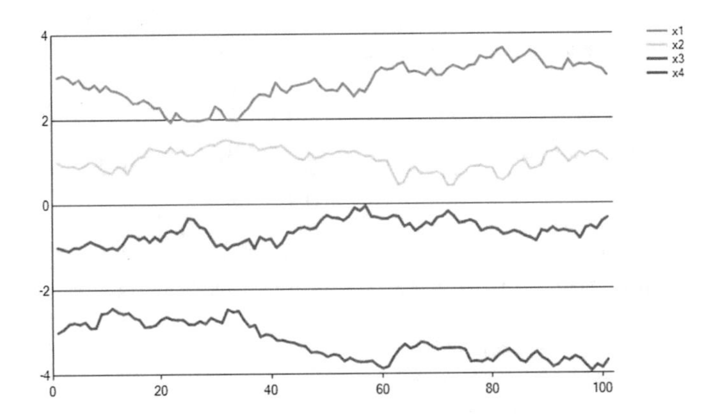

Fig. 1. Global rating of 4 alternatives: $\sigma = 1.0, c = 0.1$

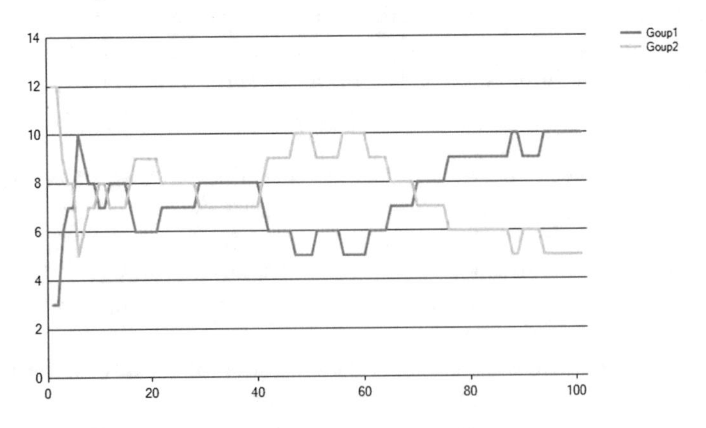

Fig. 2. Number of agents of 2 types, $\sigma = 1.0, c = 0.1$

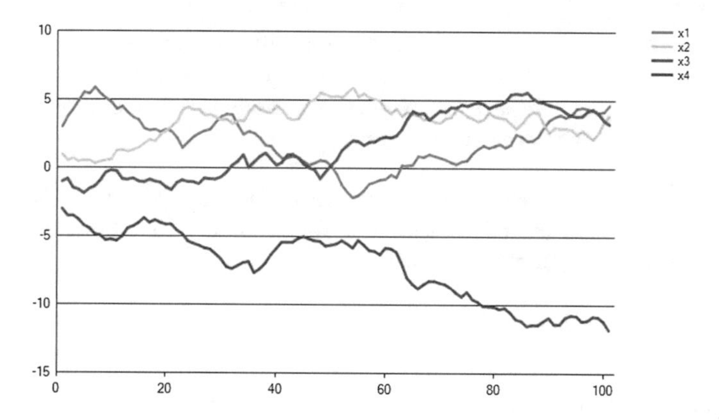

Fig. 3. Global rating of 4 alternatives: $\sigma = 3.0, c = 0.3$

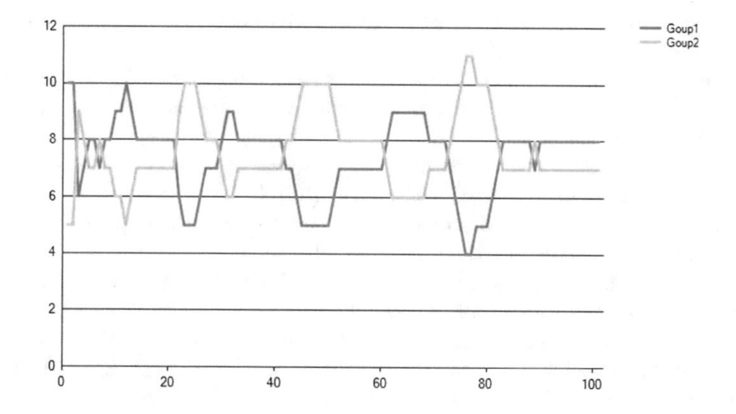

Fig. 4. Number of agents of 2 types, $\sigma = 3.0, c = 0.3$

$x_1 > x_2 > x_3 > x_4$, the rating of the alternatives has been changing a lot during the time interval from 1 to 100.

5 Conclusion

An extension of the approach from [9] deals with the problem of ranking a finite number of alternatives by pairwise comparison matrices. A dynamic system of rating for a given number of alternatives based on agent-based simulation with interactions among agents, where their PC matrices take elements from an alo-group, is presented. This system is able to simulate various business or social processes, e.g. financial market, products' demand and supply, electors' preferences of political parties, etc. A simple simulation experiment has been performed. It turns out that for small values of the parameters σ and c the global weights of alternatives remain approximately constant in the course of time. When the values of the parameters are increased then in the final time moment the vector of weights changes essentially as well as the rating of the alternatives. More simulation calculations are to be done. In the future, it will be also interesting to compare the results of various types of simulation models, e.g. additive, multiplicative, or fuzzy ones.

Acknowledgment. This paper was supported by the Ministry of Education, Youth and Sports Czech Republic within the Institutional Support for Long-term Development of a Research Organization in 2017.

References

1. Axelrod, R.: The Complexity of Cooperation: Agent-Based Models of Competition and Collaboration. Princeton University Press, Princeton (1997)
2. Anthony, P., Jennings, N.R.: Evolving bidding strategies for multiple auctions. In: Proceedings of the 15th European Conference on Artificial Intelligence, Netherlands, pp. 182–187 (2002)
3. Perzina, R., Ramik, J.: Dynamic system of rating alternatives by agents with interactions. In: Smart Innovation Systems and Technologies, vol. 58, pp. 177–185. Springer (2016)
4. Dumas, M., La Rosa, M., Mendling, J., Reijers, H.: Fundamentals of Business Process Management. Springer, Heidelberg (2013)
5. Bonabeau, E.: Agent-based modeling: methods and techniques for simulating human systems. PNAS **99**, 7280–7287 (2002)
6. Barnett, M.: Modeling and simulation in business process management (2013). http://news.bptrends.com/publicationfiles
7. Sperka, R., Spisak, M.: Transaction costs influence on the stability of financial market: agent-based simulation. J. Bus. Econ. Manage. **14**(1), S1–S12 (2013)
8. Wooldridge, M.: MultiAgent Systems: An Introduction, 2nd edn. Wiley, Chichester (2009)
9. Walsh, W.E. et al.: Analyzing complex strategic interactions in multi-agent games. In: Proceedings of the AAAI Workshop on Game-Theoretic and Decision-Theoretic Agents, pp. 1–11 (2002)
10. Cavallo, B., D'Apuzzo, L.: A general unified framework for pairwise comparison matrices in multicriteria methods. Int. J. Intell. Syst. **24**(4), 377–398 (2009)
11. Aliprantis, C.D., Burkinshaw, O.: Locally solid Riesz spaces with applications to economics. No. 105. American Mathematical Society (2003)
12. Ramik, J.: Pairwise comparison matrix with fuzzy elements on alo-group. Inf. Sci. **297**, 236–253 (2015)
13. Saaty, T.L.: Multicriteria Decision Making - The Analytical Hierarchy Process, vol. I. RWS Publications, Pittsburgh (1991)
14. Mittelhammer, R.C., Judge, G.G., Miller, D.J.: Econometric Foundations. Cambridge University Press, Cambridge (2000)

The Soft Tissue Implementation with Triangulated Mesh for Virtual Surgery System

Ruslan Akhmetsharipov(✉), Murad Khafizov, Alexey Lushnikov, and Shamil Zigantdinov

Kazan Federal University, Kazan, Russia
ahmetsharipov.ruslan@gmail.com

Abstract. In this paper, we describe specific issues arising during implementation of the virtual surgery simulator. Virtual surgery simulator is a software that provides realistic surgery experience using virtual reality technologies. We discuss the necessary requirements that a virtual surgery simulator shall meet, and suggest the possible solutions for its implementation, such as triangulated mesh for realistic rendering of soft tissue and haptic feedback. Among the supported operations with our solution for soft tissue there are cutting and stitching. Fluid dynamics are also mentioned.

Keywords: Real-time visualization · Virtual surgery · Computer simulation · Biological development · Biological dynamics · Brain

1 Introduction

The virtual surgery system was created in SIM openlab of Kazan Federal University [3]. As the presence of virtual reality in our daily lives increases, the educational system – from the kindergarten and all the way to the college – must respond to this new challenge. Naturally, medical surgical education is one of the top candidates for virtual reality technologies use, as it can save many reagents and laboratory supplies, and, as such, reduce overall monetary spending [4]. This comes with a price of possible lack of proper experience with real-life operations and reduced realism of the simulation, which in the surgery case can be downright lethal. We can, therefore, conclude that the field of virtual surgery has requirements much stricter than any other application of VR. What are those restrictions exactly?

- Highly-quality graphics, which can give the operating person a good idea of how the real thing looks;
- Realistic physics of human tissues and fluids, their proper response to various surgical procedures;
- Haptic and force feedback for user's hands according to situations in all processes;
- High level of user immersion.

As said it might be to admit, none of currently existing virtual surgery solutions are capable of matching these criteria. This leads us to the point of this paper: we propose a

G. Jezic et al. (eds.), *Agent and Multi-Agent Systems: Technology and Applications*, Smart Innovation, Systems and Technologies 74,
DOI 10.1007/978-3-319-59394-4_16

complex of technological solutions for virtual surgery simulators. Among these technologies are:

- Implementation of soft tissue physics via triangulated mesh;
- Destruction and deformation of solid mesh [5];
- Information collecting;
- Cloud storage of points necessary for rendering of multi-polygonal net;
- UV-less texturing;
- Manual interaction and haptic feedback using Glove One.

Following is description of how each of these technologies was implemented (Fig. 1).

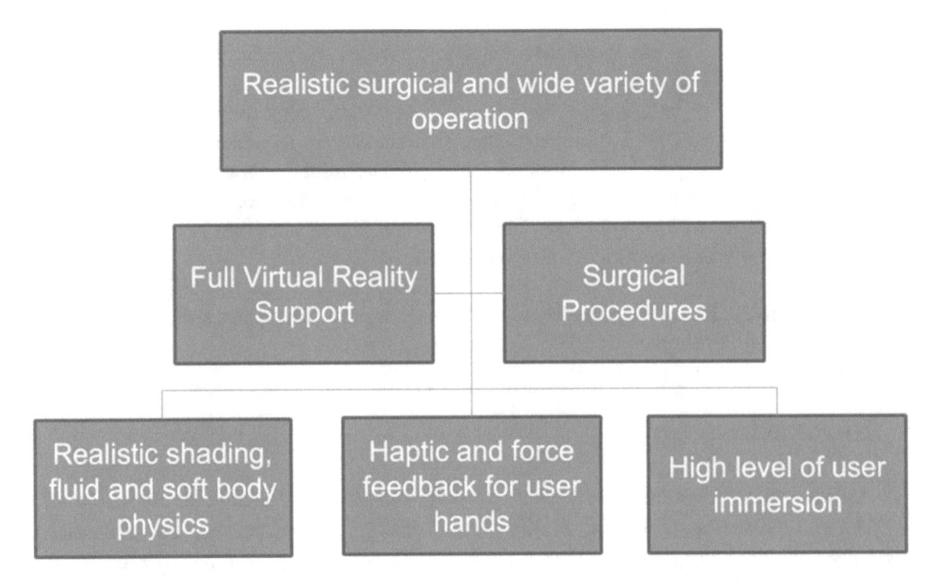

Fig. 1. Necessary things for realistic surgical operation

2 Soft Body Physics

Soft body is an object that changes its whole shape under external force. The soft body is modeled as a set of point masses (nodes) connected by perfect weightless elastic springs abiding by a variation of Hooke's law. Additional springs between nodes can be added, or the force law of the springs modified, to achieve desired effects.

A traditional rigid-body physics engine, modeling the soft-body motion using a network of multiple rigid bodies connected by constraints, and using, for example, matrix-palette skinning to generate a surface mesh for rendering, can also handle deformation. This is the approach used for deformable objects in Havok Destruction. Realistic interaction of simulated soft objects with their environment may be important for obtaining visually realistic results. Cloth self-intersection is important in some applications for acceptably realistic simulated garments. This is challenging to achieve at interactive frame rates, particularly in the case of detecting and resolving self-collisions and mutual collisions between two or more deformable objects. Detection of collisions between cloth and environmental objects with a well defined "inside" is straightforward since the system can detect unambiguously whether the cloth mesh vertices and faces are intersecting the body and resolve them accordingly. If a well defined "inside" does not exist, an "inside" may be constructed via extrusion. Mutual- or self-collisions of soft bodies defined by tetrahedra is straightforward, since it reduces to detection of collisions between solid tetrahedra (Fig. 2).

Fig. 2. Implementation flexibility for meshes in Unity3D

3 Cutting Soft Body

The cutting callback, which is recursively called to advance the cutter to the user's current position, can be further divided into five phases [1]: movement, tracking, testing, retriangulation, and update.

Movement. On this stage user moves trigger and position of trigger is projected on cuttable surface.

Tracking In an effort to reduce the time needed to test the edges of the next face for intersection, we can immediately access and test the next edge that would be crossed. By projecting dragVec onto the last edge crossed (entryEdge), we can predict if the cutter will be dragged to the edge sharing the left or right vertex of entryEdge.

Testing: On this stage application checks if user's trigger left boundaries of current face and, if this happens, go to the next stage.

Retriangulation: When trigger left boundaries of the face, application creates 2 additional vertices on the last hitted edge. Using this pair of vertices and pair vertices of last intersection current face is triangulated into 2 or 3 triangles.

Update: When cutter left current triangle and after triangulation current triangle must be updated via new information of hitted triangle.

To cut the mesh is to split the polygons located on the cutting line in half. This is accomplished by adding more edges and vertices.

Step 1. At the beginning of the process, we put a new vertex at the starting point of the cut line within the first triangle. From this vertex, we draw new edges to every other apex of this triangle, creating three new triangles as a result.

Step 2. Now, starting from this point, we start moving in the direction of the cutting tool, creating new vertices on the every edge we cross. Old edges are replaced with new ones that are connected by the new apex, and extra edges are added to avoid creating quads.

Step 3. At the end of the cut, we repeat the process of the first step: splitting the last triangle into three parts (Fig. 3).

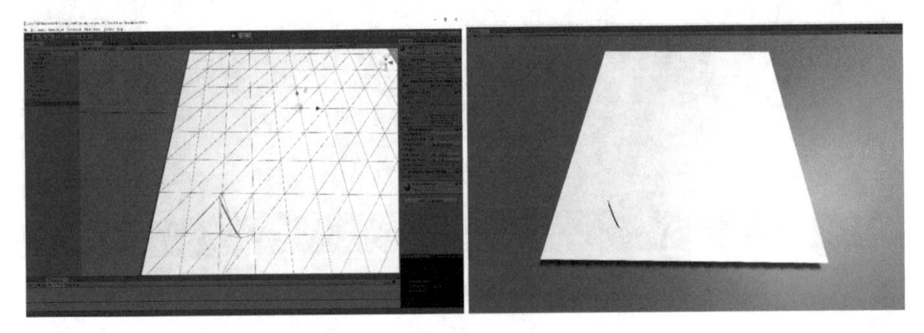

Fig. 3. The first image showing a wire frame image the cut and the second image of his appearance as the user sees

4 Stitching Triangulated Mesh and Implementing Fluids

When operation is completed within the body, it is necessary to sew up the places subjected to cutting. There are two possible ways of implementing tissue stitching.

- "Stapler-like" binding, in which specified vertices are bounded on the transformation level, passing each other information about their movement and acting in unison. In this case, we need to visibly mark the binding so it would be visible to user.
- Stitching of the mesh. This method involves connecting two different meshes together with new edges and vertices, with triangles next to the seam being replaced to absorb the edges that are both adjacent and being stitched together.

There are many mathematical models realization fluids. The main are fluids [2] based on particle system and screen space rendered fluids. In our project to display the blood system of the human body, we used the Screen Space Fluids system. This system somewhat resembles the one we have described above but demands significantly less computational power. There is no need for a separate mesh or any other game objects in SSF implementation since all of the computing is done inside of the shaders. This greatly speeds up the whole process. On other side particle based fluids much better for realistic physics simulation. In surgery simulator, we will use both of this systems for realistic blood simulation, screen space fluids for realistic look of blood and particle based fluids for blood stain surrounding objects.

5 Conclusion

Virtual surgery system has now implemented haptic feedback and processing is presented in VR model. Virtual surgery system will be used in training surgeons. Due to the multiple repetition of operations as close as possible to the real conditions, surgeons will be able to more accurately study the actual conduct of operations.

Acknowledgment. This work was funded by the subsidy of the Russian Government to support the Program of competitive growth of Kazan Federal University among world class academic centers and universities.

References

1. Bruyns, C.D., Senger, S.: Interactive cutting of 3D surface meshes. Comput. Graph. (Pergamon) **25**(4), 635–642 (2001)
2. Green, S.: Screen space fluid rendering for games. In: Game Development Conference (2010)
3. Kugurakova, V., Khafizov, M., Akhmetsharipov, R., Lushnikov, A., Galimova, D., Abramov, V., Correa, O.: Virtual surgery system with realistic visual effects and haptic interaction (2017)
4. Kugurakova, V., Talanov, M., Manakhov, N., Ivanov, D.: Anthropomorphic artificial social agent with simulated emotions and its implementation. Procedia Comput. Sci. **71**, 112–118 (2015)
5. Wu, J., Westermann, R., Dick, C.: Real-time haptic cutting of high-resolution soft tissues. Stud. Health Technol. Inf. **196**, 469–475 (2014)

Anthropic-Oriented Computing

"Thinking-Understanding" Approach in Spiking Reasoning System

Alexander Toschev[1(✉)], Max Talanov[1], and Vitaliy Kurnosov[2]

[1] Kazan Federal University, Kazan, Russia
atoschev@kpfu.ru, max.talanov@gmail.com
[2] Kazan National Research Technological University, Kazan, Russia
vitaly99@mail.ru

Abstract. In this position paper we propose the approach to use "Thinking-Understanding" architecture for the management of the real-time operated robotic system. Based on the "Robot dream" architecture, the robotic system digital input is been translated in form of "pseudo-spikes" and provided to a simulated spiking neural network, then elaborated and fed back to a robotic system as updated behavioural strategy rules. We present the reasoning rule-based system for intelligent spike processing translating spikes into software actions or hardware signals is thus specified. The reasoning is based on pattern matching mechanisms that activates critics that in their turn activates other critics or ways to think inherited from the work of Marvin Minsky "The emotion machine" [7].

Keywords: Spiking neural networks · Artificial emotions · Affective computing

1 Introduction

We suppose that the next revolutionary step in robotics will be done by autonomous systems able to make decisions and adapt themselves to complex social and dynamic environments. Unfortunately, the complexity of these environments prevents current bio plausible architectures from real-time calculations because of the lack of the computational power requested, especially in case of absence of network connectivity. Although several decades have passed between Stanford's Shakey and Honda's ASIMO, robots still have problems to perform a lot of tasks autonomously or connected through swarm protocols. DARPA DRC Trials, combining 10 tasks in human environments can help to understand the pitfalls and severe problems of autonomous robotics. For that reason, and considering the several challenges engineers are faced to, we propose computational architecture that could contribute to the improvement of adaptive skills of a robotic systems. We adopt a two-phase model in which two information processing mechanisms are combined to improve robotic processing architecture, that we call "Robot Dream" [12].

G. Jezic et al. (eds.), *Agent and Multi-Agent Systems: Technology and Applications*, Smart Innovation, Systems and Technologies 74,
DOI 10.1007/978-3-319-59394-4_17

2 The Problem

The problem of the embodient for the realistic neural simulations via spiking neural networks (SNN) seems to be not developed currently. It seems to be promising direction for the affective computing [8–10] and artificial intelligence domains to use realistic and biologically plausible simulations to reimplement psychological phenomena [5, 11, 15]. According to this approach, we could reimplement micro and macro neuro-psychological phenomena still being biorealistic. From the other perspective it seems to be promising to encapsulate bio-realistic simulations in form of robotic body, as a modern embodiment platform [2–4]. But unfortunately the computational capacity of a modern robotic system is relatively low.

2.1 Dreaming Brain Life Cycle

The "dreaming brain" is the part of the "Robot dream" project [12]. The SNN simulation of the brain that we call "dreaming brain" does not have direct connection with outer world, it uses embodiment via the robotic system, similar to the Cartesian theater described in [7]. The "dreaming brain" and the robotic system life-cycle is illustrated in Fig. 1:

1. Firstly the robotic system in "wake" phase stores the experience of every sensory channel, tagging time-frames and channels in a form of "pseudo-neuronal" activity.
2. During the direct translation the robotic system transmits to the "dreaming brain".

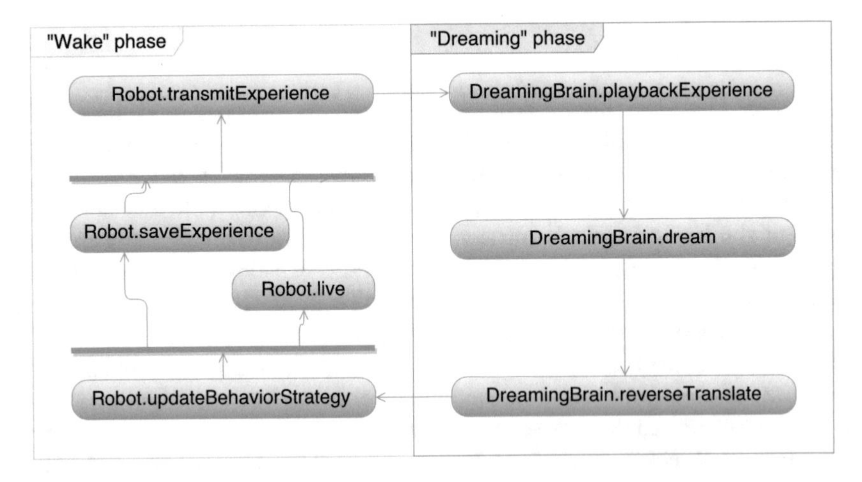

Fig. 1. The life cycle of the robotic system "wake phase" and the "dreaming brain" with direct ("wake" → "dreaming brain") and reverse ("dreaming brain" → "wake") translations.

3. The "dreaming brain" "plays back" the transmitted experience by means of activation of proper neurons, based on tagged input channel and according to stored sensory pseudo-neuronal activity.
4. The "dreaming brain" runs the simulation life-cycle updating its spiking neural network. There are two options: to use phases of the mammalian dream or to use wake simulation.
5. During the reverse translation the "dreaming brain" runs the translation the updated structure of sNN into rules of behavioral strategies of the robotic system.

3 Our Idea

We propose rule based system which stands on TU framework [14]. As implementation of the rule based system runnable real-time or semi real-time we propose to use TU framework described in [13] that utilize "Critic-Selector-Way to think" (T3) model described in [14] to manage the robotic system. The T3 "Critic-Selector-Way to think" triplet inherited from works of Marvin Minsky [6, 7] and provides an option to evaluate incoming sensory data over the stored knowledge in the knowledge base. The inbound information of the TU framework [14] is textual now we propose to use spikes as the representation of inbound and processing information of the TU framework. The TU framework is based on probabilistic rules and despite neural networks uses logical reasoning with the 6-levels of mental activity and T3 over spikes [7].

3.1 Probabilistic Critics

Spikes trigger several critics, that start inbound information processing in parallel on several levels of mental activity. Critics are grouped in contexts based on their level of mental activity and semantics of the processing information (audial, visual, tactile). The activation of one critic of the context increases the probability of triggering of corresponding critics of the context. This way every critic is a temporal probabilistic predicate that contains set of rules that are evaluated not only over the incoming information, but over current system state and context of recently processed information.

The overall architecture is presented on the picture below. Incoming spikes activates critics by evaluating attached to the critic set of rules. See section General Workflow for overall information. Activate critic runs the selector, which return another critic or way-to-think.

We propose to evaluate logic rules via NARS [16] or PLN [1]. PLN and NARS are probabilistic reasoning systems. Once final score will be produced system specific critic activates and way-to-think checks Knowledge Base for activation limit for this specific spike. Knowledge base is used for internal learning and rule storing. System also takes in account the number of the incoming spikes. Several spikes will be only accumulated and $n + 1$ spikes trigger a process. For example, robotic's ultra sonic sensor detects approaching to the obstacle. It starts generation of the frequency and produces next data:

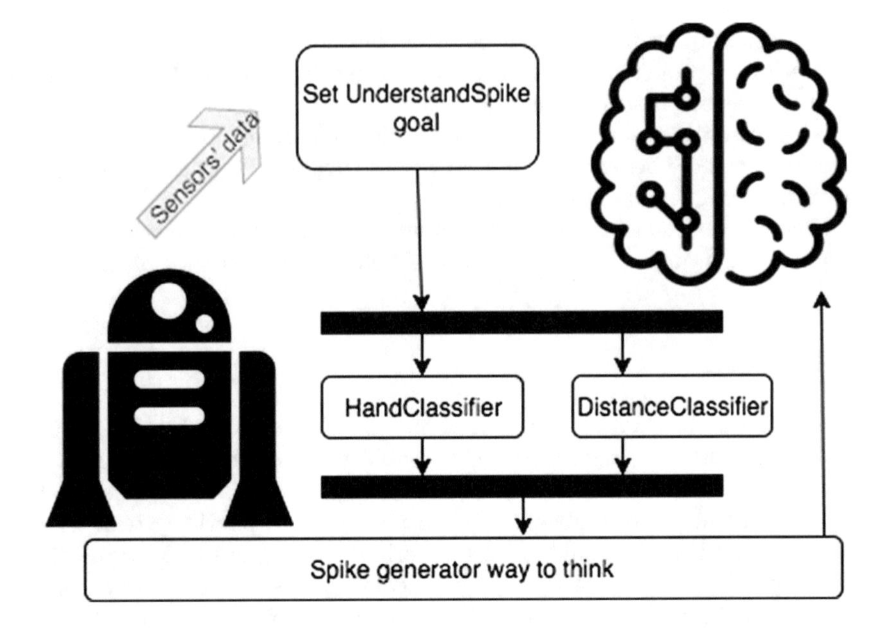

Fig. 2. Reasoning lifecycle.

```
{channel: 0,
data: [ {2313, 2017-02-02 13:33:21:122121},
                  {3223, 2017-02-02 13:33:21:122121},
                  {123,2017-02-02 13:33:21:122121},
                  {12331,2017-02-02 13:33:21:122121}]};
```

These data processed by TU. TU set UnderstandSpike goal. Critics attached to goal are activated: HandClassifier, DistanceClassifier critics. They have rules which applied over the incoming data. See Fig. 2 for visualisation.

```
{
 MATCH Goal
 WHERE UnderstandSpike
 RETURN true
}
=>
{
 START SpikeGeneratorWayToThink.Way2Think,
 ContextChangedWayToThink.Way2Think
}
```

3.2 Way to Think

Way to think according TU design is a "worker" — component which performs actual modification of a data in the Knowledge base or raises action. Way to think can be used as different actions:

1. Append score for specific type of incoming data. For example, if incoming data is not enough to run the send the signal it will be accumulated. There are some kind of necessary limit for activation;
2. Send signal to hardware;
3. Do nothing and wait for incoming data.

A way to think also can run a workflow that could trigger the hardware controller of the robotic system. Way to think actually modify data or perform action. If translation is direct way to think start generation of the spikes into NEUCOGAR. For example, new type of way to think added to the system – Spike generator Way to think. It's goal to generate neuron spikes according to classified data. For example, input data:

```
family: hand, foot, e.t.c.
period: in second
```

Output will be call to NEUCOGAR to generate spikes.

4 General Workflow

Input data for the system represented by two types according to workflow path. See Fig. 3.

1. Robotic sensors' data. Represented by the JSON array

```
{channel: 0,
data: [ {2313, 2017-02-02 13:33:21:122121},
{3223, 2017-02-02 13:33:21:122121},
{123,2017-02-02 13:33:21:122121},
{12331,2017-02-02 13:33:21:122121}]};
```

2. NEUCOGAR spike processing result – neuron spikes.

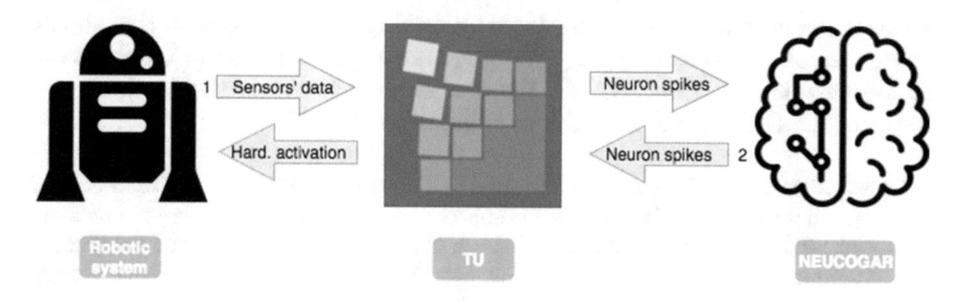

Fig. 3. General workflow.

In general there are 2 type of workflow direct – from robot to NEUCOGAR and from NEUCOGAR to robot. Below the sample workflow for 1st and 2nd path.

4.1 Direct Translation: 1st Path

1. Sensors' data hits TU;
2. TU creates inbound context for data based on its type and attributes: channel, sensor's data, time, previously processed information;
3. TU starts processing the context;
4. Several Critics are activated based on resulting probability of their rules;
5. Several data will be only accumulated in the system state and raise spike generation only after accumulated count will be enough to trigger;
6. The critic activates way to think;
7. Way to think generates neuron spikes for NEUCOGAR.

4.2 Reverse Translation: 2nd Path

1. Spike hits TU;
2. TU creates inbound context for this spike based on its type and attributes: origin, time, activated number of neurons, mediators, previously processed information;
3. TU starts processing the spike;
4. Several Critics are activated based on resulting probability of their rules;
5. Several spikes will be only accumulated in the system state and raise motor reaction only after accumulated count will be enough to trigger;
6. The critic activates way to think;
7. Way to think generates data for controllers.

5 Related Work

This paper is a part of Robot Dream project [12]. Related works include description and testing of the TU framework [13] for automatic incident processing. Due to architecture it can be simply extend.

6 Conclusion

In this paper we have provided the intro for spiking reasoning system that should be used as robotic control system. We propose to use TU framework and reasoning systems like PLN and NARS. We provide description of the workflow and usages of the Critic, Selector, Way to think from the TU.

Acknowledgments. This work was funded by the subsidy allocated to Kazan Federal University for the state assignment in the sphere of scientific activities, grant agreement no. 1.2368.2017; by the subsidy of the Russian Government to support the Program of competitive growth of Kazan Federal University among world class academic centers and universities.

References

1. Goetzel, B., Mathew, I.: Probabilistic Logic Networks: A Comprehensive Conceptual, Mathematical and Computational Framework for Uncertain Inference. Springer, New York (2008)
2. Haikonen, P.O.: Example: an experimental robot with the HCA. In: Consciousness and Robot Sentience, pp. 203–224. World Scientific (2012)
3. Haikonen, P.O.A.: Consciousness and sentient robots. Int. J. Mach. Conscious. **5**(1), 11–26 (2013). http://www.worldscientific.com/doi/abs/10.1142/S1793843013400027
4. Haikonen, P.O.A.: Yes and no: match/mismatch function in cognitive robots. Cogn. Comput. **6**(2), 158–163 (2014). http://link.springer.com/10.1007/s12559-013-9234-z
5. Leukhin, A., Talanov, M., Sozutov, I., Vallverdú, J., Toschev, A.: Simulation of a fear-like state on a model of dopamine system of rat brain. Adv. Intell. Syst. Comput. **449**, 121–126 (2016)
6. Minsky, M.: The Society of Mind. Simon & Schuster, New York (1988)
7. Minsky, M.: The Emotion Machine: Commonsense Thinking, Artificial Intelligence, and the Future of the Human Mind. Simon & Schuster, New York (2007)
8. Picard, R.W.: Affective Computing. Massachusets Institute of Technology (1997)
9. Picard, R.W.: What does it mean for a computer to "have" emotions? In: Trappl, R., Petta, P., Payr, S. (eds.) Emotions in Humans and Artifacts (2001)
10. Picard, R.W., Vyzas, E., Healey, J.: Toward machine emotional intelligence: analysis of affective physiological state. IEEE Trans. Pattern Anal. Mach. Intell. Graph Algorithms Comput. Vis. **23**(10), 1175–1191 (2001)
11. Talanov, M., Vallverdu, J., Distefano, S., Mazzara, M., Delhibabu, R.: Neuromodulating cognitive architecture: towards biomimetic emotional AI. In: 2015 IEEE 29th International Conference on Advanced Information Networking and Applications, vol. 2015, pp. 587–592. IEEE, March 2015
12. Tchitchigin, A., Talanov, M., Safina, L., Mazzara, M.: Robot Dream, pp. 291–298. Springer, Cham (2016)
13. Toschev, A.: Thinking model and machine understanding in automated user request processing. CEUR Workshop Proceedings, vol. 1297, pp. 224–226 (2014)
14. Toschev, A., Talanov, M.: Thinking lifecycle as an implementation of machine understanding in software maintenance automation domain. Smart Innovation Syst. Technol. **38**, 301–310 (2015)
15. Vallverdú, J., Talanov, M., Distefano, S., Mazzara, M., Tchitchigin, A., Nurgaliev, I.: A cognitive architecture for the implementation of emotions in computing systems. Biologically Inspired Cogn. Architect. **15**, 34–40 (2015)
16. Wang, P.: Non-Axiomatic Logic: A Model of Intelligent Reasoning. World Scientific, Singapore (2013)

Pseudorehearsal in Value Function Approximation

Vladimir Marochko, Leonard Johard, and Manuel Mazzara(✉)

Innopolis University, Universitetskaya street 1, 420500 Innopolis, Russia
m.mazzara@innopolis.ru

Abstract. Catastrophic forgetting is of special importance in reinforcement learning, as the data distribution is generally non-stationary over time. We study and compare several pseudorehearsal approaches for Q-learning with function approximation in a pole balancing task. We have found that pseudorehearsal seems to assist learning even in such very simple problems, given proper initialization of the rehearsal parameters.

Keywords: Reinforcement learning · Rehearsal · Pseudorehearsal · Catastrophic forgetting

1 Introduction

Reinforcement learning is a more general problem formulation than the commonly used supervised learning framework. As such, it can be applied to a wider range of problems. It is also a more difficult problem to optimize for, as the feedback is more limited.

Reinforcement learning covers space of prediction and control problems in partially unknown space with unknown behavior. The agent should explore the environment and find optimal actions it has to perform to reach the goal. It is used in cases when the optimal policy is unknown so there is no way to train agents using supervised learning algorithms. The basic idea can be described by the metaphor of a player starting an unknown game and, after a number of turns, he/she receives a message stating that he/she has lost or won. After a number of games he/she will figure out how to act to win as often as possible.

1.1 Supervised Agents in Reinforcement Learning Problems

In order to solve the practical problems that can be assumed to be approximately Markov decision process (MDP) [1], like robot's navigation, playing chess or trading on stock exchange, we can use a value function approximation to speed up the learning process [2,3]. The weakness of this approach is that convergence is not guaranteed if the MDP approximation is incorrect, or in cases where the inputs are continuous, and which necessitates non-linear function approximation [4]. Furthermore if we have continuous outputs the value approximation

G. Jezic et al. (eds.), *Agent and Multi-Agent Systems: Technology and Applications*, Smart Innovation, Systems and Technologies 74,
DOI 10.1007/978-3-319-59394-4_18

needs to be combined with an additional optimization technique, such as REINFORCE [5], in order to search for optimal outputs. If the outputs are discrete, a simple maximization is sufficient and allows us to make use of the Q-learning framework [6,7].

Although a value function simplifies the learning problem by effectively converting a reinforcement learning problem to a supervised learning problem through the use of bootstrapping [8], it is still a more difficult problem than conventional supervised learning. One of these additional difficulties is that the policy-dependent rewards introduces a concept drift [9]. This introduces the risk of unstable oscillations, but is generally solvable at the cost of slower learning if the learning rate is set sufficiently small in the beginning. This ill conditioning of the problem has been considered one of the main challenges of reinforcement learning.

1.2 Catastrophic Forgetting

A different problem which recently got more attention is catastrophic forgetting [10]. This problem is most commonly described in an online unsupervised Hebbian learning task, where the ability to retrieve previously stored patterns is lost as we update weights in the training of new patterns. This catastrophic forgetting of the original patterns takes place even when parameter space is more than sufficient to store both sets of patterns and is a consequence of the limited mixing of input objects.

Simply alternating between the new and old patterns group with a sufficiently low learning rate would in theory solve the problem, but this has a potential impact on the convergence rate and requires explicit memorization of all training patterns. In order to minimize the effect on convergence rate, we would like to maximize the mixing of the presented inputs.

2 Catastrophic Forgetting in Reinforcement Learning

The online nature of reinforcement learning means that catastrophic forgetting is a key bottleneck. There are two principal non-sharpening approaches to the catastrophic forgetting problem suggested in literature: rehearsal and pseudorehearsal. In addition, other methods based on sparse representations [11,12] have been used less frequently. This latter approach has a theoretical downside in its negative impact on the ability to generalize, but has shown at least mixed results and is possible to use in conjunction with the rehearsal methods.

2.1 Rehearsal Approaches

The first and most straight-forward principal approach for mitigating catastrophic forgetting is rehearsal [13,14]. A rehearsal strategy simply stores a subset of all previous experiences in a buffer. When a new pattern is presented, this pattern is combined with several patterns from the buffer in order to form a learning batch with good mixing. There are several possible heuristics for selecting patterns for rehearsal.

The importance of catastrophic forgetting in reinforcement learning was identified early. Lin introduced the term Experience Replay [15] for referring to the use of rehearsal strategies in the reinforcement learning setting. Such rehearsal has shown very promising results in robotics [16] and on more complex environments, such as Deep Q-learning for playing Atari games [17].

2.2 Pseudorehearsal Approaches

A second principal approach to solving catastrophic forgetting is pseudorehearsal [18], which does not require explicit storage of patterns. Instead, it uses a two-step process where generative models are learnt alongside with the main task. These generative models create pseudopatterns, which are combined in batches with real patterns for training the agent.

An interesting questions is whether these generated approximations of the real data are sufficiently accurate in practice to reduce forgetting. Remarkably, even extremely crude generative models have proven highly effective. In the original work in this area by [18], pure noise fed to the network was able to almost completely eliminate catastrophic interference. The argument of the authors was that, although the input is completely random, the activation distributions in deeper levels of the network will be representative of the learnt input data.

An analytical approach by Frean and Robins [19] in single perceptrons suggest an alternative explanation for the surprising efficiency of random pseudopatterns. They suggest that the pseudopatterns approximate the mean of the input. Training on this mean of the input leads to decorrelation of the input patterns, which in high dimensional inputs makes the different patterns' weight updates orthogonal to each other. In addition, they demonstrated that using this mean directly was at least efficient as generating pseudopatterns. Further work in this direction was done in a thesis by Goodrich [20], where some of these results where expanded to multilayer perceptrons.

Regardless of the reason for such networks, pseudorehearsal methods have been demonstrated to significantly decrease and almost completely eliminate the catastrophic forgetting in unsupervised learning [18], supervised learning [13] and reinforcement learning [21]. It is interesting to note that the results of Baddeley suggest that the widely studied ill conditioning might not be the main bottleneck of reinforcement learning after all. Instead, their results indicate that the catastrophic forgetting is the main bottleneck for reinforcement learning problems.

Pseudorehearsal Algorithms. For testing pseudorehearsal approach we used two different pseudorehearsal types and the online learning with one backpropagation step as an example of learning without pseudorehearsal. One algorithm is based on correcting of the weight updates, other is a batch-backpropagation learning.

The first pseudorehearsal algorithm is the one used by Frean and Robins [19] with a simplified weighting equation and changed for non-linear neural network inner assignments. The idea behind algorithm is to generate pseudoset, feed it through the network and save activations on each neuron for every pseudoexample. Then the agent is learned online, but when the real example fed to the network we use equation

$$\Delta w_i = err_{b_i} \frac{1}{pr} \sum_{j=1}^{pr} \frac{b_i x_{ij} \cdot x_{ij} - x_{ij} x_{ij} \cdot b_i}{b_i \cdot b_i x_{ij} \cdot x_{ij} - b_i \cdot x_{ij} b_i \cdot x_{ij}}$$

to update w_i - weights at the ith layer, where err_{b_i} - vector backpropagation errors of the learned example at the ith layer, pr - size of pseudoset, b_i - vector of activations of the learned example when fed forward through the network and x_{ij} is vector of activations of the jth pseudoset on the ith layer.

The second one - is straight-forward using pseudosets in batch backpropagation learning - we generate set of pseudoexamples, feed it through the network, save the network outputs as the targets and then create a matrix of feature vectors where first vector is real example, others are pseudovectors, and matrix of targets where first vector is target for the real example and others are saved earlier networks outputs on pseudoset, then each time we learn agent on the whole set.

2.3 Biological Forgetting

An interesting particular case of catastrophic forgetting problem is learning in the human brain. Dual network models were initially inspired by biological learning [22]. As a consequence of promising experimental results of such networks, pseudorehearsal was indeed found to be the most plausible explanation for the otherwise cryptic need for dual learning systems in the brain [23].

More biologically detailed extensions of these models have recently been explored by Hattori, where they again showed excellent improvements on the ability to store information [24].

The pseudorehearsal approach also contributes to the urgent need for new biological plasticity rules in large scale neurosimulation and especially for their developmental varieties (e.g. BioDynaMo [25]). Real full scale brain simulation is approaching, but we are still lacking even a basic understanding of the role dreams play in the learning process. This despite the fact that sleep stages are of considerable length and evident in even the simplest of biological neural networks.

3 Experimental Design

We will reevaluate the analytic results of Frean and Robins [19] in a real reinforcement learning task. We evaluate and compare two algorithms for pseudorehearsal on a pole balancing task using Q-learning with function approximation. Further, we will study the effect of sparsity in fulfilling the requirement for a high dimensional input space these algorithms relied on. Our agent is a classic Q-learning agent with ϵ-greedy policy using a feed-forward-backpropagation neural network as function approximator, discounted factor of agent is 0.9. The environments used for training is the single-pole balancing cart.

Observation. Two different observations are used for experimental comparison. The first observation type given to the agents constitute a fully observable MDP and includes current position, velocity, acceleration of the cart, as well as the current angle, angular velocity and angular acceleration of the pole. The second observation type make the problem partially observable - here the agent knows only cart's position and pole's angle. If the cart reaches the end of track or the pole falls for angle more than predefined pole failing angle - the game is lost and the agent is gained negative reward for this task.

We represented the observations as a feature vector by two different methods. The first method was a representation of input values where the i-th observation value sets the $2 * i$-th feature if it is positive or on $2 * i + 1$-th if it is negative. The second method was to use sparse unary vectors where feature vector is concatenation of parameter vectors, similar to a table. Each of parameter vectors consisted of elements associated with discrete values inside the range possible for each parameter - $[-20; 20]$ for linear parameters, $[-60; 60]$ for angular. All the elements of the vector were set to zeros except two - the element associated with the rounded value of the parameter was set to one, and the next element is set to the fractional part of this parameter.

Performance Metric. Agent tries to balance pole or poles as long as it can for 5000 tries, for two sets of parameters we also made an averaged variants where 10 iterations of this 5000 tries are averaged to make sure that convergence tendency is reproducible and not a set of random successful moves. We also have results for fully random policy to compare with.

Parameter Settings. The task was repeated with different sizes of pseudoitem batches, with different numbers of iterations between reinitialization of the pseudosets and with different learning rates. The learning rates used were 0.1, 0.01 and 0.001. The discount factor was set to 0.9. The sizes of pseudosets were 10, 30, 50 and 100 pseudoitems, respectively. We resample a new set of pseudoitems after every 1, 10 or 100 runs. Parameters are chosen to define influence of size of the pseudoset and frequency of it's reinitialisation on learning. We try to cover a wide range without trying all the possible values. And we decrease parameters in close to geometrical progression to see if the influence is logarithmic. For 30 and 50 item pseudorehearsal batches we also tried 30 and 50 reinitialisation gaps to increase coverage.

Performance Metrics. We did not stop learning after an agent reaches satisfactory result, because the continued learning contains cases of catastrophic forgetting which we would like to explore. As we had a very short learning time during which the performance increases followed by long row of tries with unstable behavior, we evaluate the efficiency of different approaches by measuring mean and median number of steps per try for each approach and compare these two numbers.

Mean > *median* indicates that some agent's tries were highly effective and agent balanced pole for a long time, while the most of runs were weak, so no convergence occurred or the influence of catastrophic interference is too high to handle needed weights for a long time. *Mean* < *median* shows that agent successfully converge to some optimal policy, and it's policy is stable but some tries are failed so bad that it affected the whole picture, so in this case we can see successful learning, strong influence of catastrophic forgetting and successful avoidance of this influence. *Mean* ≈ *meadian* means that catastrophic forgetting and its avoidance has nearly the same influence. Results were averaged over ten runs when training times allowed and other cases we presents results over single runs (Fig. 1).

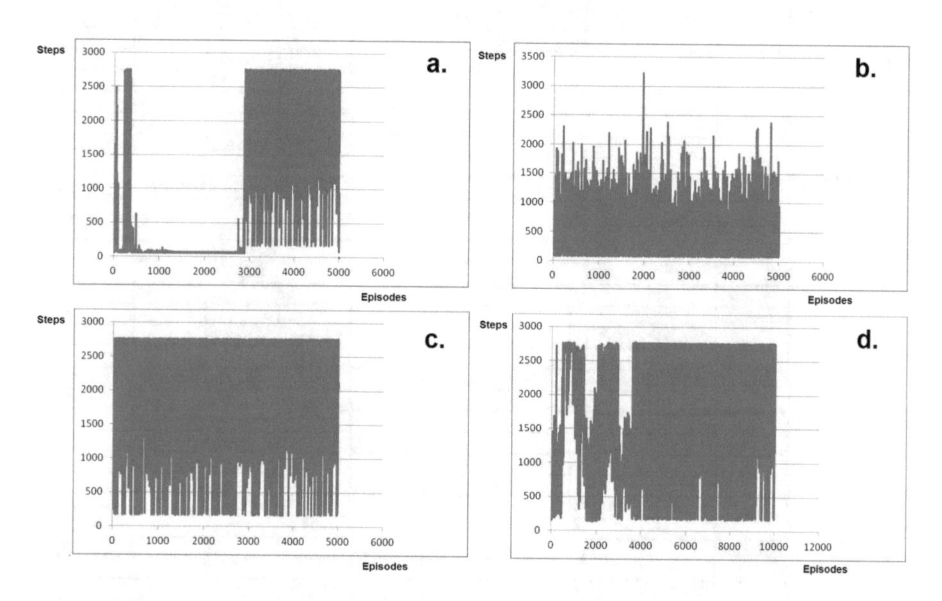

Fig. 1. Examples: a. median ≪ mean; b. median < mean; c. median > mean; d. median ≈ mean

3.1 Results

Results for all observations show different learning for different cases, while averaged approach and comparison with the random agent assure us that the learning has place and in case without pseudorehearsal it depends on learning rate mostly. For the agents using pseudorehearsal learning depends on sets of parameters - learning rate, pseudoset size and relearning gap. Some of them can make agent

Table 1. MDP results overall

Random		mean	217.72							
		median	206							
Learning rate	Fully observable 0.1					Fully observable 0.01				
No PR1	767.7612					525.7532				
	460					398				
No PR2	653.255					485.9627				
	355					370				
Frean-Robins	pseudoset	10	30	50	100	pseudoset	10	30	50	100
	relearn					relearn				
	1	88.4	479			1	53.8746	1425.314		
		84	339				53	1209		
	10	198.9974	260	1784	288.7504	10	81.161	61.0631	955.4878	107.554
		119	211	1938.5	210		54	54	1060	81
	30		573			30		460.5927		
			442					415		
	50			655		50			168.7599	
				276.5					66	
	100	722	413		1982.6	100	95.4113	272.1958		178.9365
		297	217		2058		82	107		113
batch	pseudoset	10	30	50	100	pseudoset	10	30	50	100
	relearn					relearn				
	1	181.9208	234.434			1	66.2465	388.2868		
		53	68				53	244.5		
	10	578.7582	159.062	562.722	92.5414	10	113.2239	73.4251	525.2255	1853.865
		119	89	324.5	55		53	53	400	1873
	30		1085.417			30		64.727		
			725					53		
	50			503.2082		50			2032.241	
				91					2142.5	
	100	389.2236	747.8532		56.3434	100	85.1528	178.7005		1065.519
		155	383		53		56	92		651
	Fully observable 0.001					Fully observable 0.1 averaged				
	316.564					1561.47782				
	169					1569.55				
	528.7564					1561.47782				
	398					1569.55				
Frean-Robins	pseudoset	10	30	50	100	pseudoset	10	30	50	100
	relearn					relearn				
	1	117.7718	251.8404			1	608.9561	433.9956		
		92	163				609.25	423.5		
	10	84.1524	110.2418	73.33	694.8546	10	460.5518	859.295	705.3484	309.2548
		56	86	79	663.5		467.15	862	696.7	296.85
	30		549.4772			30		447.0809	284.7513	
			434					439.7	262.4	
	50			54.258	309.9588	50				
				53	76					
	100	513.2786	74.9174			100	423.9774	677.3001		481.9075
		383	75				413.65	679.85		480.55
batch	pseudoset	10	30	50	100	pseudoset	10	30	50	100
	relearn					relearn				
	1	409.2142	1114.044			1	447.7925			
		299	818				420.1			
	10	369.9044	290.125	76.6654	253.8526	10	500.5673	542.4561		471.3174
		238	84	56	103		487.55	516.05		454.15
	30		410.3088			30		492.2506		
			325					465.35		
	50			267.8876		50				
				191						
	100	72.6846	177.4432		278.7542	100	381.1586			481.4641
		54	73		180		363.9			463.9

to balance pole for significantly larger time than a random run, while the others perform the same, or worse, or even worse than a random agent.

For both used metrics we discovered a roughly bell-shaped graph of dependencies for each of used learning rates and for each of used techniques. All pseudorehearsal approaches have different sets of parameters providing optimal learning, a suboptimal and worse - in case of parameters a little different from the optimal ones. Further away from optimal parameters agents started to diverge, e.g. tried to drop the pole about four times faster than if it would use random policy, etc.

Both Frean-Robins and batch approaches perform similarly in for this observation, but their optimal parameters differ for the same learning rates. All this results are summarized in Table 1 for MDPs and Table 2 for partially observable MDPs - POMDPs.

Table 2. POMDP results overall

Learning rate	Fully observable, sparse vector, 0.01					Partially observable, 0.01					Partially observable, 0.01, averaged by 10 runs				
No PR1 mean	233.8168					1856.8124					617				
median	54					1961					610				
No PR2 mean	271.5718					1422.7302					880				
median	54					1348.5					880				
Frean-Robins	pseudoset	10	30	50	100	pseudoset	10	30	50	100	pseudoset	10	30	50	100
	relearn					relearn					relearn				
	1	178.994	182.0386			1	2027.527	74.9296			1	411.4776	830.6546		
		162	162				2128	54				411.35	833.85		
	10	197.9846	212.739	87.5092	186.0228	10	2034.459	62.829	2006.305	2065.93	10	972.3696	560.9303	950.2528	413.022
		180	177	76	175		2124.5	62	2075.5	2064.5		981.6	560	952.8	413.05
	30		246.1306			30		1297.337			30		1084.086		
			140					1449					1091.3		
	50			127.0978	109.941	50			2034.181		50			730.0647	
				107	91				2106.5					735.2	
	100	124.3928	156.9248			100	1588.98	58.0668		1174.671	100	644.3332	376.2169		885.305
		105	143				1876	53		1209		644.6	377.7		885.85
batch	pseudoset	10	30	50	100	pseudoset	10	30	50	100	pseudoset	10	30	50	100
	relearn					relearn					relearn				
	1	61.3802	151.1766			1	72.9094	151.1766			1	491.1629	289.2957		
		56	53				53	53				474.7	235.05		
	10	64.867	61.3236	78.0882	79.2608	10	143.184	270.6204	81.5922	701.0954	10	298.6029	581.6812	431.0811	696.3091
		59	54	64	60		53	59	53	312		273.4	570.5	424.3	669.8
	30		413.445			30		79.1648			30		485.2032		
			317					62					484.95		
	50					50			1356.528		50			331.9525	
									1313					320.7	
	100	66.006	85.9946		129.6376	100	361.9262	947.519		53.362	100	359.6483	469.9983		403.4266
		56	76		98		266	79		53		327.25	443.6		376.45

3.2 MDP

For the cases of fully observable MDP, where agent knows anything about the current state of the pole cart and the pole the learning time when the agent's performance goes from some initial random to the final one is very short, agent quickly converges at some number of steps it can balance and most of its next moves holds around this result with some deviations, sometimes very large, caused by catastrophic forgetting. If some learning case makes agent significantly change its behaviour - change is as quick as initial learning and on graph looks like immediate change of performance (Fig. 2).

3.3 POMDP

Partial observability tends to suffer less from forgetting, possibly because each part of the smaller space are more frequently visited. Pseudorehearsal has a more

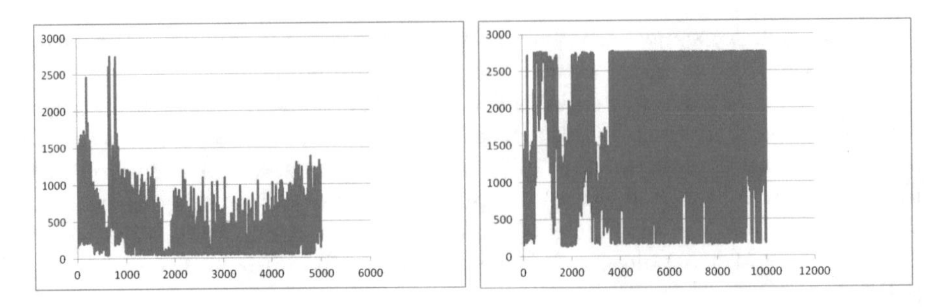

Fig. 2. Left plot - learning rate = 0.001, batch learning, pseudoset size = 10, relearning gap = 10; right - learning rate = 0.01, pseudoset size = 100, relearning gap = 100

significant impact here: although it shares the same optimal-suboptimal-worst sets of parameters, optimal ones further decrease the number of runs needed to learn and decrease influence of the catastrophic forgetting: if the agent in current set of parameters doesn't diverges to the worst possible case - it's median for all runs is always higher then the mean, indicating a relatively stable behaviour after training. On the other hand, agents in POMDPs agents that can't converge change their policy more frequently than agents in MDPs, and while the agent in fully observable MDP has a minor chance to reach good performance after it reached a suboptimal solution, agent in POMDP easily changes its policy both ways (Fig. 3).

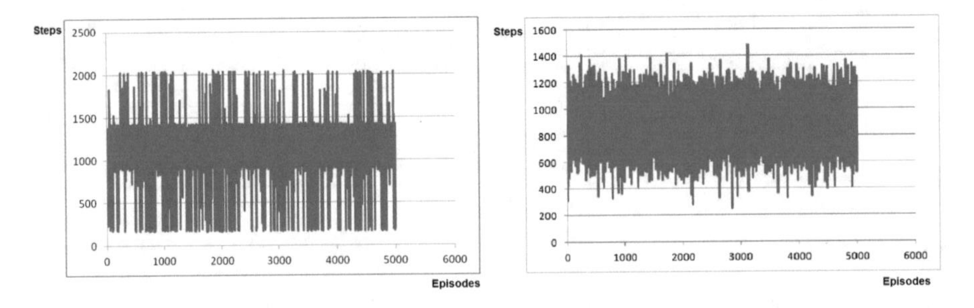

Fig. 3. Results of running in POMDP with pseudoset size 100 and relearning gap 100, single run on the left plot and averaged by 10 runs on the right one

The pseudorehearsal approach taken from Frean and Robins decreases this serious context switches and all the agents using this type of pseudorehearsal show nearly the same performance during the all runs, holding around some value with occasional deviations, results much better or worse can be met, but they are rare compared to the results close to this mean (Fig. 4).

Stability is maintained for all sets of parameters, while the mean value can differ. Different effect is caused by batch-backpropagation learning using pseudosets: picture of agent's performance is the same as in learning without

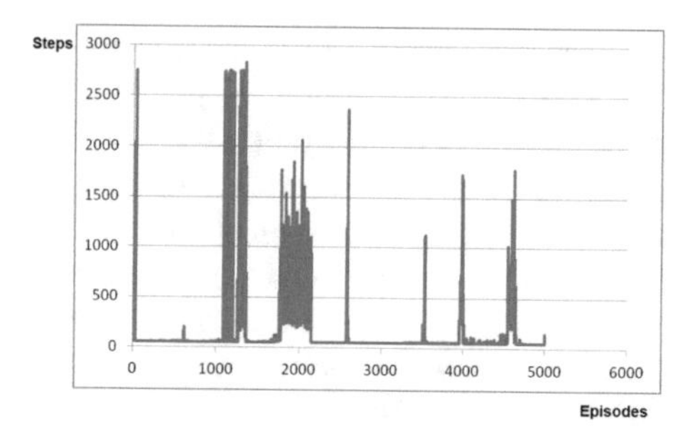

Fig. 4. Example of diverged agent changing policy: batch pseudorehearsal, pseudoset size = 30, no relearning gap - new pseudoset generated after each run

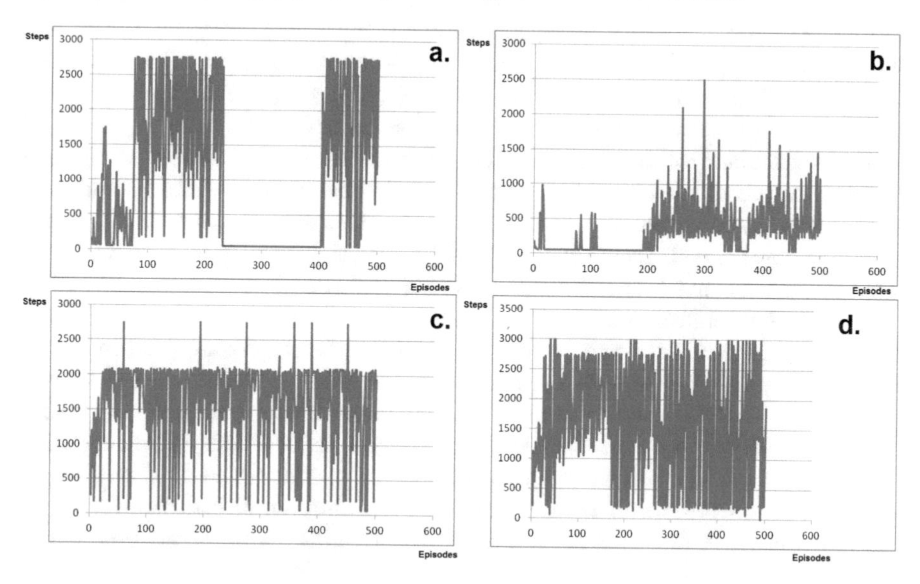

Fig. 5. a. 500 first steps for an agent without pseudorehearsal (seed 1); b. 500 first steps for an agent without pseudorehearsal (seed 2); c. one agent with Frean-Robins' pseudorehearsal with pseudoset size = 10 and relearning gap = 100; d. one agent with batch pseudorehearsal with pseudoset size = 100 and relearning gap = 10

pseudorehearsal, but efficiency switching occurs only after reinitialization of the pseudoitem vector. While same pseudorehearsal parameters may lead to different agent's behaviour, the efficiency of each set of parameters evaluated by averaging for ten iterations shows that some sets make agent to increase it's performance during the time, while the others do not (Fig. 5).

4 Discussion and Conclusions

The experiment has shown us that pseudorehearsal can deal with catastrophic interference, but it has its own effects which in some cases cause divergence that worsen performance, so this tool should be used carefully and the parameters - learning rate, pseudoset size and relearning frequency have to be chosen properly to guarantee high performance on the current task.

For the fully observable MDPs pseudorehearsal decreases influence of the catastrophic forgetting if the optimal parameters for the task are known. In the best cases, optimal performance was reached quickly with pseudorehearsal, but the further parameters from the optimal, the worse performance was. In the case if modeling this environment might be too complex - optimal parameters can be defined only empirically before starting learning, which may be unacceptable if the cost of mistake is high.

For partially observable environments all the problems met by fully observable ones remain the same and some additional effects were noted: agent's policy doesn't only converge, diverge or stay random, but also converge to some value with the majority of tries having results in a some range around this value, and strongly deviating runs are more rare and separated by wide gaps of convergence. Another notable effect of the pseudorehearsal in POMDP agents with both pseudorehearsal cases is a significant decrease of the number of steps to converge to the number of steps needed by fully observable agent. If an agent in fully observable environment can converge it converges at about 20–30 runs as with pseudorehearsal, so without, if agent in partially observable environment can converge it converges at 100–250 runs without pseudorehearsal and at 10–30 runs with pseudorehearsal.

Pseudorehearsal is known to be a powerful tool for improving performance of supervised learning agents. We have shown that it can be useful to assist learning even in relatively quickly mixing continuous reinforcement learning tasks, if parameters are chosen correctly. Pseudorehearsal reduces this forgetting effect and maintains stable solutions for longer. While pseudorehearsal may strongly improve agent's performance and accelerate learning, empirical defining of the optimal pseudoset size and relearning gap is required. One of possible extension of this research would be exploration of mathematical way to figure out this parameters. We will also explore new, more complex reinforcement learning challenges and try more advanced dual network generation of pseudoexamples.

References

1. Murphy, K.P.: A survey of pomdp solution techniques. Environment **2**, X3 (2000)
2. Johard, L., Ruffaldi, E.: A connectionist actor-critic algorithm for faster learning and biological plausibility. In: 2014 IEEE International Conference on Robotics and Automation (ICRA), pp. 3903–3909. IEEE (2014)
3. Sutton, R.S.: Learning to predict by the methods of temporal differences. Mach. Learn. **3**(1), 9–44 (1988)
4. Tsitsiklis, J.N., Van Roy, B.: An analysis of temporal-difference learning with function approximation. IEEE Trans. Autom. Control **42**(5), 674–690 (1997)
5. Williams, R.J.: Simple statistical gradient-following algorithms for connectionist reinforcement learning. Mach. Learn. **8**(3–4), 229–256 (1992)
6. Watkins, C.J., Dayan, P.: Q-learning. Mach. Learn. **8**(3–4), 279–292 (1992)

7. Geist, M., Pietquin, O.: A brief survey of parametric value function approximation. Rapport interne, Supélec (2010)
8. Sutton, R.S., Barto, A.G.: Reinforcement Learning: An Introduction, vol. 1, no. 1. MIT Press, Cambridge (1998)
9. Gama, J., Sebastião, R., Rodrigues, P.P.: On evaluating stream learning algorithms. Mach. Learn. **90**(3), 317–346 (2013)
10. McCloskey, M., Cohen, N.J.: Catastrophic interference in connectionist networks: the sequential learning problem. Psychol. Learn. Motiv. **24**, 109–165 (1989)
11. French, R.M.: Semi-distributed representations and catastrophic forgetting in connectionist networks. Connection Sci. **4**(3–4), 365–377 (1992)
12. Coop, R., Mishtal, A., Arel, I.: Ensemble learning in fixed expansion layer networks for mitigating catastrophic forgetting. IEEE Trans. Neural Netw. Learn. Syst. **24**(10), 1623–1634 (2013)
13. Ratcliff, R.: Connectionist models of recognition memory: constraints imposed by learning and forgetting functions. Psychol. Rev. **97**(2), 285 (1990)
14. Hinton, G.E., Plaut, D.C.: Using fast weights to deblur old memories. In: Proceedings of the Ninth Annual Conference of the Cognitive Science Society, pp. 177–186 (1987)
15. Lin, L.-J.: Reinforcement learning for robots using neural networks. Technical report, DTIC Document (1993)
16. Adam, S., Busoniu, L., Babuska, R.: Experience replay for real-time reinforcement learning control. IEEE Trans. Syst. Man Cybern. Part C (Appl. Rev.) **42**(2), 201–212 (2012)
17. Mnih, V., Kavukcuoglu, K., Silver, D., Graves, A., Antonoglou, I., Wierstra, D., Riedmiller, M.: Playing atari with deep reinforcement learning. arXiv preprint arXiv:1312.5602 (2013)
18. Robins, A.: Catastrophic forgetting, rehearsal and pseudorehearsal. Connection Sci. **7**(2), 123–146 (1995)
19. Frean, M., Robins, A.: Catastrophic forgetting in simple networks: an analysis of the pseudorehearsal solution. Netw. Comput. Neural Syst. **10**(3), 227–236 (1999)
20. Goodrich, B.F.: Neuron clustering for mitigating catastrophic forgetting in supervised and reinforcement learning. Ph.D. dissertation, University of Tennessee (2015)
21. Baddeley, B.: Reinforcement learning in continuous time and space: interference and not ill conditioning is the main problem when using distributed function approximators. IEEE Trans. Syst. Man Cybern. Part B (Cybern.) **38**(4), 950–956 (2008)
22. McClelland, J.L., McNaughton, B.L., O'Reilly, R.C.: Why there are complementary learning systems in the hippocampus and neocortex: insights from the successes and failures of connectionist models of learning and memory. Psychol. Rev. **102**(3), 419 (1995)
23. Robins, A., McCallum, S.: The consolidation of learning during sleep: comparing the pseudorehearsal and unlearning accounts. Neural Netw. **12**(7), 1191–1206 (1999)
24. Hattori, M.: A biologically inspired dual-network memory model for reduction of catastrophic forgetting. Neurocomputing **134**, 262–268 (2014)
25. Breitwieser, L., Bauer, R., Meglio, A.D., Johard, L., Kaiser, M., Manca, M., Mazzara, M., Rademakers, F., Talanov, M.: The biodynamo project: creating a platform for large-scale reproducible biological simulations. In: 4th Workshop on Sustainable Software for Science: Practice and Experiences (WSSSPE4) (2016)

Finding Correlations Between Driver Stress and Traffic Accidents: An Experimental Study

Margarita Pavlovskaya, Ruslan Gaisin, and Rustem Dautov(✉)

Higher Institute of Information Technology and Information Systems (ITIS), Kazan Federal University (KFU), Kazan, Russia
ritapavlovskaya@yandex.ru, rusgaisin@gmail.com, rdautov@it.kfu.ru

Abstract. As the number of people getting injured or killed on the roads is constantly growing, it is crucial to identify and prevent potential factors causing traffic accidents. This paper focuses on one of such factors – namely, the drivers' stress, which is known to be one of the main causes of traffic accidents, and timely detection of such situations becomes an important challenge. The paper aims to find a potential correlation between the driver stress when riding through a specific urban location and the recorded history of traffic accidents in that specific location. If proven, such a correlation can help to prevent traffic accidents and re-design urban spaces in a safer manner. To achieve this goal, the paper combines cross-disciplinary techniques from Computer Science and Physiology to measure drivers' stress levels using physiological sensors during city rides, and match these experimental results against a map of previously recorded traffic accidents. As a result, the conducted study indicates that the correlation indeed exists, and measuring drivers' stress levels using physiological sensors is a promising approach to minimise the amount of traffic accidents.

Keywords: Stress detection · Physiological sensors · Traffic accident · Cube of emotions

1 Introduction

Traffic accidents cause considerable damage to national economies worldwide. As of 2013, the World Health Organisation (WHO) reported 1.25 million people killed in traffic accidents with another 20–50 million people sustaining non-fatal injuries as a result of road traffic collisions or crashes [1]. Factors leading to this alarming statistics are manifold, among which driver stress is seen as one of the main problems. Driver stress typically occurs as a result of curtain conditions or circumstances experienced by the driver before or during the ride. These stressful conditions represent a curtain emotional and psychological workload on the driver, which can be quantified. As suggested by [9], there are four levels of driver stress – namely, *no stress*, *low stress*, *medium stress*, and *high stress*. Arguably, the latter three need to be avoided and prevented, as they are most likely to

G. Jezic et al. (eds.), *Agent and Multi-Agent Systems: Technology and Applications*, Smart Innovation, Systems and Technologies 74,
DOI 10.1007/978-3-319-59394-4_19

lead to a traffic accident. In this light, it becomes a promising attempt to investigate a potential dependency between the drivers' stress levels and occurrences of traffic accidents in areas, where increased stress levels are observed. Understanding of changes in the driver's stress level can potentially lead to an ability to predict drivers' further actions, their general ability to drive, and, eventually, to determine their readiness for a trip.

Measuring of such bodily reactions to surrounding conditions is nowadays possible by means of a variety of physiological sensors, each of which serves to measure a certain metric of a human body, such as heart beat rate, body temperature, or blood sugar level. Accordingly, this paper presents an experimental study, which aims to find a potential correlation between the drivers' stress levels and occurrences of traffic accidents. The presented cross-disciplinary research effort lies at the intersection of Computer Science and Physiology, and is expected to contribute to both fields. The presented experiments have been conducted in the city of Kazan – a modern megapolis in the European part of Russia, characterised by a high number of cars, urban public transport, and, unfortunately, an increased rate of road traffic accidents.

The rest of the paper is organised as follows. Section 2 briefs the reader on related research works, and puts forward the main hypothesis. Section 3 describes the experiment methodology, and explains its individual steps. Section 4 summarises the collected information and analyses the experimental results. Section 5 concludes the paper.

2 Related Work and Hypothesis

The concept of finding a dependency between human emotions and psychological states and the rate of traffic accidents, potentially caused by these emotions, is not completely novel. A relevant experimental study has been previously conducted by Healey and Picard [4], who used physiological sensors to collect data and analyse it in two ways. First, they used features from 5-minute intervals of data during the rest, highway, and city driving conditions to distinguish three levels of driver stress with an accuracy of over 97% across multiple drivers and driving days [4]. In their second research paper [5], the authors compared continuous features, calculated at 1-second intervals throughout the whole duration of a drive, with a metric of observable 'stressors' created by independent coders from videotapes. As a result, both experiments demonstrated that skin conductivity parameters and heart rate are most closely correlated with the drivers' stress level. Based on these findings, the authors argued that it is possible to implement a mechanism, which will be able to determine the level of the driver stress, and, thus, can be used to create non-critical embedded car systems, which would automatically control various functions – e.g. the radio may become quieter, the navigation system may notify of critical moments, the phone may decline/redirect incoming calls, etc. Moreover, once implemented and integrated into on-board safety systems, such a solution would detect critical levels of the driver stress, and take preventive actions accordingly to avoid or minimise potential consequences of dangerous – that is, stressful – situations, which can lead to a traffic accident.

This fundamental concept of measuring physiological states of a human body to detect stressful conditions, underpins the experimental study to be presented in this paper. Accordingly, this paper puts forward the following hypothesis – given that the driver stress level can be measured by a variety of physiological sensors, is there really a correlation between an increased stress level and a possibility of a traffic accident, occurring on a given road segment? In other words, this paper tries to find out whether it is possible to predict a traffic accidents on a very specific location by continuously measuring a driver's stress level while driving through that specific location. In order to prove this hypothesis, the paper first presents a methodology and equipment used to conduct experiments, and then proceeds with the actual experiments.

3 Methodology and Experiment Setup

As already introduced, by using physiological sensors it is possible to measure electric signals, produced by the human body in real-world driving conditions – this is an important factor to achieve maximum accuracy when measuring the drivers' daily psychological workload, as opposed to a laboratory testbed, where simulation techniques are frequently used.

Driving a car in a megapolis is frequently fraught with *stress*. The traffic situation in a big city requires continuous attention and concentration, which may result in constant nervous tension. Psychologists have found three most powerful stress factors for a person, which dominate the other minor factors [8]:

- *Uncertainty*: often road signs and the real traffic directions do not match. Inexperienced drivers cannot find the right route in the maze of streets and road junctions. Many drivers get lost when driving through busy intersections.
- *Suddenness*: traffic conditions nowadays have become difficult to predict. Unexpected obstacles and interferences, sudden noises, unplanned road works, emerging jams are just a few examples.
- *Responsibility*: understanding the fact of having a drivers license leads to an increased awareness and rand associated responsibility to drive safely for passengers and other drivers.

Taken together, these factors lead to an increased stress level which can be seen as the human body's reaction to any demand that exceeds the human adaptive capacity. The stress affects attention, anxiety, working memory, and perceptual-motor performance. In this light, driving is not an exception, and is widely influenced by various 'stressors' on an every-day basis [3].

As suggested by the study of emotional receptivity and neurotransmitters by Lövheim [7], monoamines (i.e. serotonin, dopamine, and noradrenaline) have a considerable impact on the people's mood, emotions and behavior. These relationships are depicted by the so-called 'cube of emotions' [7] (see Fig. 1). The cube represents a three-dimensional model for monoamine neurotransmitters and

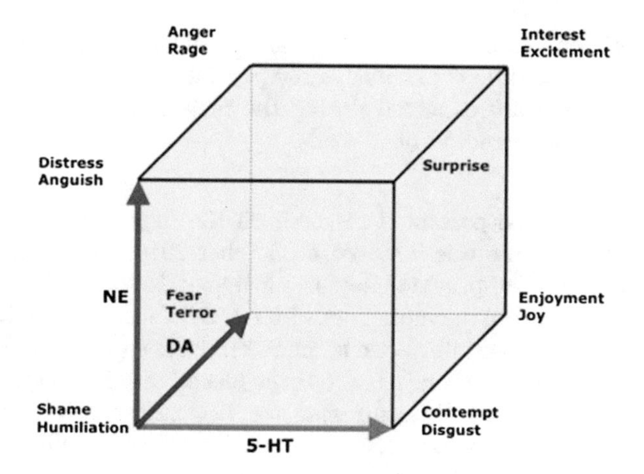

Fig. 1. A three-dimensional model for emotions and monoamine neurotransmitters (excerpted from [7]).

emotions. In this model, the monoamine systems are represented as orthogonal axes and the eight basic emotions are placed at each of the eight possible extreme values, represented as vertices of the cube.

Accordingly, it has been identified that each of the monoamines is involved in different aspects of human emotions and behavior. More specifically, a combination of increased levels of noradrenaline and dopamine and a decreased level of serotonin leads to an increased stress level. This key pattern has been used in the presented study to detect drivers' stress levels.

3.1 Setup of the Experiments

Necessary prerequisites for conducting the experiment were the following: a valid driver's license, a route map and instructions for the driver, a camera, physiological sensors and a laptop. More details on the main four steps of the proposed methodology to conduct the experiment are described below.

Preparing Physiological Sensors. The following physiological sensors were used for the experiment: OpenBCI 32bit Board and OpenBCI Daisy Module[1] were used to sample brain activity (i.e. electroencephalogram - EEG), muscle activity (i.e. electromyogram - EMG), and heart activity (i.e. electrocardiogram - ECG), and a corresponding 3D-printed headset[2] was used to attach the OpenBCI system to the human head. The system is equipped with wireless interfaces to communicate to both stationary PCs and mobile devices. Using these sensors, the data are collected during the driver's rest and during the drive on various

[1] http://shop.openbci.com/products/openbci-16-channel-r-d-kit.

[2] https://irenevigueguix.wordpress.com/2016/07/15/behind-the-ultracortex-mark-iii-novaand-supernova/.

roads with different traffic conditions. It assumed that different parts of the driving route indicate different signals of the physiological sensors. In particular, lower levels of stress are expected during the rest, and higher (or sometimes critical) levels are expected on busy roads.

Preparing Data. The presented research on the dependence of the driver's stress and road accidents was initiated in October 2016, when city traffic data were first requested. The provided datasets included 70 spreadsheets for 13 districts of the city of Kazan, including data from CCTV cameras, which indicated regulated and unregulated intersections, small blocks or segments of busy streets. Additionally, statistical information about the places with increased rate of traffic accidents was provided. By comparing the two datasets, it was possible to find a correlation – that is, potentially stressful segments of urban roads with an increased number of traffic accidents.

Determining the Driving Route Based on the Map of Accidents. The experimental route consists of multiple roads, highways and by-streets in the city of Kazan, and is based on historical data about traffic accidents, which were requested from the local traffic police. Based on the acquired information, it was possible to come up with a 'map of accidents' – that is, a map of most dangerous road sections. Using this map, a suitable route (i.e. route passing through the maximum number of dangerous spots) was selected.

Conducting the Experiments and Collecting Results. Six drivers volunteered to participate in the experiment, and agreed to be recorded by a camera and equipped with physiological sensors during 20 rides. Sensors were attached, so that the drivers would not be distracted from the road and would drive carefully. Before the drive, all the drivers were shown the route map and instructed accordingly. All the trips were taken during the morning rush hours, i.e. from 7.30 AM to 9.30 AM, and on weekends, when the congestion on the roads was reduced by 1.5–2 times, compared to working days. As a result of these experiments, data were collected in the form of plotted graphs, containing measurements of the physiological sensors.

4 Recognising Emotions

The first step was to analyse the collected data. Since all the results were summarised in the form of graphs, this enabled enabled clear analysis of the emotions, experienced by the drivers. Please note that in all three diagrams below the horizontal axis represents time, and the vertical line corresponds to a moment when the experiments were commenced.

Analysis of the Data: Distress. *Distress* is an aversive state, in which a person is unable to adapt completely to 'stressors' and the associated stress, thus demonstrating maladaptive behavior. As it can be seen on Fig. 2, all the graphs have an ascending trend.

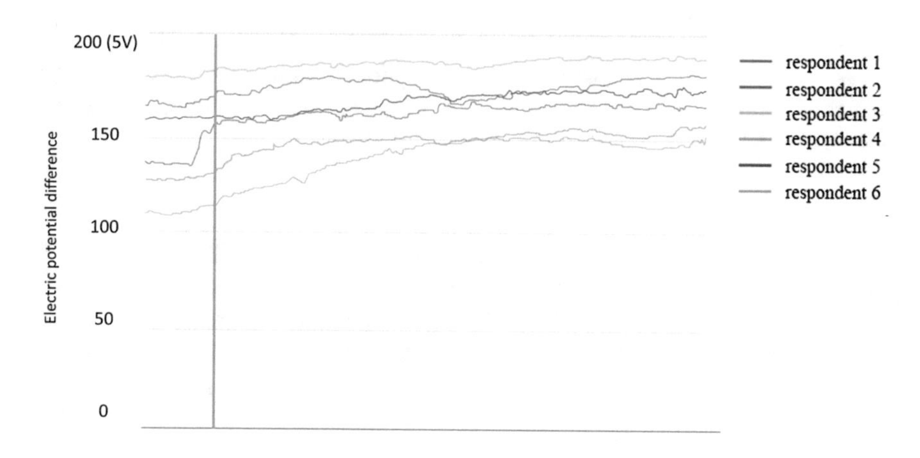

Fig. 2. Recognising distress.

Analysis of the Data: Fear. *Fear* is the internal negative emotional state, caused by a real (or anticipated) threat. It is a natural human reaction to danger. As it follows from Fig. 3, it is possible to detect the moments when the fear appears – i.e. these are emotional states (caused by the driver's conditions during the experiment), which are represented by sudden peaks in the sensor measurements.

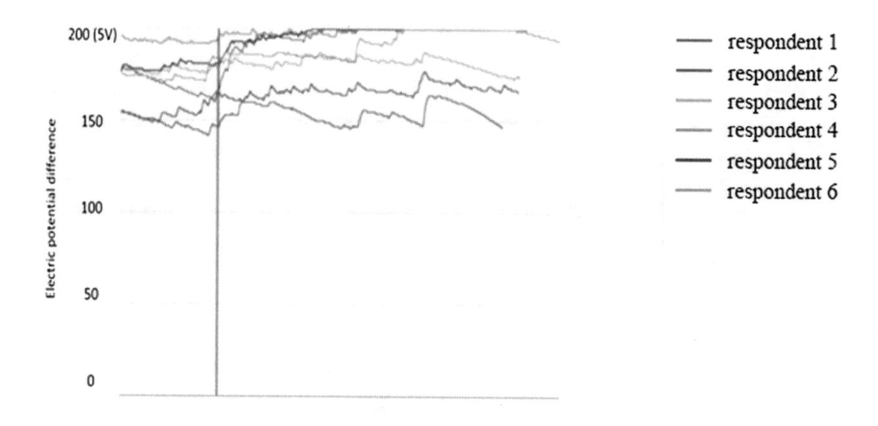

Fig. 3. Recognising fear.

To determine the certain emotions, the following formula was used to calculate the biggest fluctuation:

$$max|x_i - x_{i-k}|, i = \overline{1, n} \tag{1}$$

where x_1 is the arithmetic mean value of the original sample (i.e. before the start of the experiment), x_2 is the arithmetic average value of the final sample (i.e. the last seconds of a certain segment of the experiment). This is the maximum value of the fear – the peak value on the graph. The maximum standard deviation of fluctuations is calculated according to:

$$max(|x_i - x_{i-k}|), i = \overline{1, n} \tag{2}$$

Assuming that this value is the peak of an emotional state, which represents the maximum expression of fear, this numerical model allows to distribute drivers on a scale of expressed emotions. Also, during their trips through the city, drivers were asked on how they were feeling themselves. By comparing the data on the graphs and the answers of the subjects, it follows that the proposed method accurately determines the fear emotions and has the potential to be used to verify the numerical simulation.

Analysis of the Data: Disgust. *Disgust* is a negative emotion, which appears in deep hostility and plays an important role in human self-protective function [2]. Disgust can be divided into primary (i.e. unconscious psychic reaction of the body) and secondary (i.e. moral or psychological). The graphs depicted in Fig. 4 enable analysis of the disgust emotions experienced by the drivers. As it follows, in all experiments there is a decreasing trend. For the numerical model of disgust, the following hypothesis was applied:

$$\overline{x_1} - \overline{x_2} \tag{3}$$

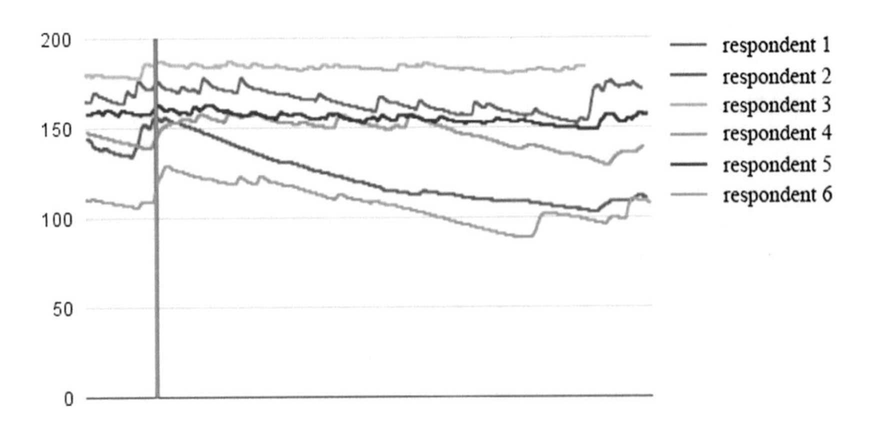

Fig. 4. Recognising disgust.

where x_1 is the arithmetic mean value of the original sample (i.e. before the start of the experiment), x_2 is the arithmetic average value of the final sample (i.e. the last seconds of a certain segment of the experiment). This value characterises the overall level of disgust. It is assumed that the higher the value, the higher the level of disgust. The indicator that characterises fluctuation intensity of the graph is the following:

$$\sum_{i=1}^{n} = |x_i - x_{i-k}| \tag{4}$$

where n is the total number of values, x_i is the current value, and x_{1-k} is the value with a step of size k. This value characterises variability of the emotional state. It is assumed that the greater the value, the more stable the emotional state. The standard deviation is the following:

$$S = \sqrt{\frac{1}{n}\sum_{i=1}^{n}(x_i - \overline{x})^2} \tag{5}$$

where n is the total number of values, x_i is the current value, and $\overline{x}$ is an arithmetic mean value. This value represents the total change of the emotional state. It is assumed that the greater the value, the higher the level of disgust.

4.1 Finding Correlations

The final step of the experiment was to check whether increased emotional levels were observed while drivers were passing through a potentially dangerous junction, aiming to prove the initial hypothesis. The second factor for proving the hypothesis was personal observation – i.e. it was possible to match sensor readings with personal observations. For example, whenever sensors indicated low stress levels, the drivers behaved calmly and drove confidently, and vice versa. Upon completion of all experiments, all sensor readings and personal notes on the drivers' behaviour were compared to the 'map of accidents'. As a result, the following conditions should be met at the same time for a given location on the 'map of accidents' in order to claim that the hypothesis is valid:

- An increased level of noradrenaline (i.e. emotions of distress).
- An increased level of dopamine (i.e. emotions of fear).
- A decreased level of serotonin (i.e. emotions of disgust).
- The driver reports on a stressful state.
- The current geo-location of the driver is within one of the dangerous areas.

Taking all these factors into account, the following two observations can be drawn in this respect:

1. It appeared that the proposed hypothesis is valid to quite a certain extent. Considering all similarities and differences, it was possible to calculate the

correlation coefficient, which equaled to about 0.72, which means that in 72% of the cases, the drivers demonstrated an increased stress level when passing through a dangerous junction (i.e. a junction with an increased rate of previously recorded accidents). There were also a few cases, when drivers rode relatively calmly through a dangerous area. This can be explained by the fact that accidents in these areas in the past were caused by loss of concentration, sleepiness, not being focused, etc. – all those activities, which are not detected by high stress levels, but nevertheless may still lead to traffic accidents. It is worth noting that these indicators are correlated with weather conditions, road surface conditions, the state of the vehicle, as well as general physical condition of the driver.

2. The second observation was the following. Indeed, in most locations marked as dangerous – i.e. where crashes, assaults on pedestrians, cyclists or obstacles were reported – sensors indicated a deviation from the usual emotional state of the drivers. However, these deviations were not necessarily represented as increased stress levels, but as decreased levels as well. This means that at some locations the drivers were over-relaxed and self-confident to such an extent, that would loose concentration and, as a result, potentially cause a traffic accident. At these points the drivers were driving in an 'unconscious' manner, not paying much attention to the surrounding context and assuming the road was completely safe to drive.

Taking into account both observations, the initial hypothesis has to be slightly refined – it can be claimed that there appears a correlation not only between the rate of traffic accidents on a specific spot and increased levels of the drivers' stress levels, but with decreased levels as well. In other words, a considerable (both positive and negative) deviation from the usual emotional state may potentially lead to a traffic accident.

It is also worth noting that the described approach relied on a two-factor identification of the stress state - i.e. using the physiological sensors and interviewing drivers on their emotions during the drive. As a potential way of introducing a more reliable and precise three-factor identification, measuring the level of *cortisol* in the human saliva is seen as a next step for further research. Cortisol is known to be released in response to the external stress factors, and is already used to detect stress levels [6].

5 Conclusion

The paper demonstrated that the monitoring and detection of the drivers' stress levels during real urban drives is feasible using physiological sensors. Using the 'cube of emotions', it was possible to detect three main emotions – i.e. fear, disgust and distress – which are dependent on changes in the levels of the three neurotransmitters – i.e. dopamine, noradrenaline, and serotonin. Collected experimental data served to build mathematical models of the three emotions, according to which the stress state is directly proportional to the level of dopamine and noradrenaline, and indirectly proportional to the level of serotonin. This makes it possible to determine the state and psycho-emotional deviations of the driver during the drive. Also, the paper aimed to prove the hypothesis that increased stress levels are observed while riding through dangerous road segments. The hypothesis appeared to be partially true, as it was also observed that decreased stress levels also have an impact on the potential occurrence of a traffic accident at a given spot.

Acknowledgments. This work was funded by the subsidy allocated to Kazan Federal University for the state assignment in the sphere of scientific activities.

References

1. Global Health Observatory (GHO) data. http://www.who.int/gho/road_safety/mortality/en/, Accessed 02 Jan 2017
2. Gaisin, R., Gaisina, K.: Mathematical modelling of human fear and disgust emotional reactions based on skin surface electric potential changes. In: Proceedings of the 2017 International Conference on Artificial Life and Robotics (ICAROB 2017), pp. 331–333 (2017)
3. Ge, Y., Weina, Q., Jiang, C., Feng, D., Sun, X., Zhang, K.: The effect of stress and personality on dangerous driving behavior among chinese drivers. Accid. Anal. Prev. **73**, 34–40 (2014)
4. Healey, J., Picard, R.: Smartcar: detecting driver stress. In: 2000 Proceedings of the 15th International Conference on Pattern Recognition, vol. 4, pp. 218–221. IEEE (2000)
5. Healey, J.A., Picard, R.W.: Detecting stress during real-world driving tasks using physiological sensors. IEEE Trans. Intell. Transp. Syst. **6**(2), 156–166 (2005)
6. Hellhammer, D.H., Wüst, S., Kudielka, B.M.: Salivary cortisol as a biomarker in stress research. Psychoneuroendocrinology **34**(2), 163–171 (2009)
7. Lövheim, H.: A new three-dimensional model for emotions and monoamine neurotransmitters. Med. Hypotheses **78**(2), 341–348 (2012)
8. Reason, J.: Human error: models and management. West. J. Med. **172**(6), 393 (2000)
9. Rigas, G., Goletsis, Y., Fotiadis, D.I.: Real-time driver's stress event detection. IEEE Trans. Intell. Transp. Syst. **13**(1), 221–234 (2012)

Towards Robot Fall Detection and Management for Russian Humanoid AR-601

Evgeni Magid and Artur Sagitov(✉)

Intelligent Robotic Systems Laboratory,
Higher Institute for Information Technology and Information Systems (ITIS),
Kazan Federal University, Kazan, Russian Federation
{magid, sagitov}@it.kfu.ru

Abstract. While interacting in a human environment, a fall is the main threat to safety and successful operation of humanoid robots, and thus it is critical to explore ways to detect and manage an unavoidable fall of humanoid robots. Even assuming perfect bipedal walking strategies and algorithms, there exist several unexpected factors, which can threaten existing balance of a humanoid robot. These include such issues as power failure, robot component failures, communication disruptions and failures, sudden forces applied to the robot externally as well as internally generated exceed torques etc. As progress in a humanoid robotics continues, robots attain more autonomy and enter realistic human environments, they will inevitably encounter such factors more frequently. Undesirable fall might cause serious physical damage to a human user, to a robot and to surrounding environment. In this paper, we present a brief review of strategies that include algorithms for fall prediction, avoidance, and damage control of small-size and human-size humanoids, which will be further implemented for Russian humanoid robot AR-601.

Keywords: Robot control · Humanoid robots · Safety · Humanoid robot fall · Safe fall · Fall prediction · AR-601

1 Introduction

Humanoid robotics is still considered a rather a young research field with many research challenges. While industrial robots are being widely used in manufacturing and their technology have reached high level of maturity with a variety of robots available from different manufacturers, only a few humanoid robots are currently commercially available. Most of full-size humanoids are built on customer request and come with a high price tag. Even though these humanoids share majority of their components (e.g. harmonic drive gears, controllers, sensors etc.), the systems differ significantly.

Humanoid robot locomotion is an extremely challenging research field as keeping stability during standing straight and locomotion is a necessary requirement for all applications of such robots. To address this issue problem of dynamically stable biped locomotion received significant attention over the last decades with some promising results [26–29].

G. Jezic et al. (eds.), *Agent and Multi-Agent Systems: Technology and Applications*, Smart Innovation, Systems and Technologies 74,
DOI 10.1007/978-3-319-59394-4_20

Biped humanoid robots have several advantages over wheeled mobile robots as they can step over obstacles and go up and down stairs. On the other hand, a bipedal robot has a major disadvantage – it may fall over and then get seriously damaged, injure people or destroy surrounding objects. Today this, together with high power consumption, is the most significant barrier for practical application of humanoid robots. Therefore, humanoid robots could not be entirely integrated in our society as everyday human assistants and companions unless this problem is solved.

There is a number of general approaches to solve this issue. First approach, Robot Hardware Design, deals with robot hardware; it concentrates on engineering robots' hardware in a way that it could survive a fall over due to resistant materials usage, shock-absorbing structures, etc. Second approach, Fall Detection, emphasizes importance of detecting when a fall is imminent in order to avoid such situations. Third approach, Fall Management, proposes a special fall sequence for reducing robot body damage or damage to objects in vicinity.

Fall avoidance strategies are an attempt to reduce fall frequency. When a fall does occur, fall damage control strategies could potentially minimize robot damage and/or damage of its environment. As humanoid robots are generally heavy, robot fall generally results in its serious damage or causes various damages to an object that is hit by the robot. Particularly, as an upper body of a humanoid robot is positioned relatively high when the robot moves, stands or performs some operations in straight vertical pose, the damage is likely to be very substantial. Therefore, it is desirable for a humanoid robot to minimize any damage, which the robot or an object hit thereby suffers when the robot turns over. While falling motion control reduces the robot damage, landing impact may still damage its parts if experiments are repeated over again to reevaluate control parameters. Therefore, most researchers have to substitute real experiments with simulations in order to reexamine and refine the control.

Fall damage minimizing received an interest in human biomechanics and have been extensively studied for the past decades [5–8]. Even though biomechanics experience and contribution are valuable for a fall detection and management, and provides significant insights, we should be aware of the limits that are imposed by differences between biology and mechatronics, which emphasize that biomechanics results could not be directly applied to robots. For example, behavior of humans during a fall evolved with an instinct to save high-value regions of the body first, firstly protecting a head, a frontal face, or any limb that was ever injured previously. This may be applicable with corrections for a humanoid robot to protect an area of essential circuitry. There are also differences in the materials of the body and motor control between a robot and humans, which makes direct transfer impractical.

The rest of the paper is organized as follows. Section 2 presents fall detection approaches for small size and human size humanoids. Section 3 deals with fall management. Section 4 discusses our future work proposal on fall detection and prevention for AR-601. Finally, we conclude in Sect. 5.

2 Fall Detection

The main objective of fall detection algorithms is to discriminate between fall events and normal activities of a robot. If during an operation, a humanoid robot suffers a strong disturbance under external force or torque, and its controller generates a correction motion that cannot actually be completed on time or performed in general, then the robot might fall even though the stabilizing control keeps operating.

To detect a fall Renner et al. in [9] used attitude sensors and indicators that triggered integrated into a control system robot reflexes. They estimated model parameters from an ideal gait sequence and used deviations that were calculated from the robot sensory data in real time as an instability indicator, which triggers recovery process.

Ogata et al. [10] proposed fall prediction methods that are based on a predicted Zero Moment Point (ZMP). The predicted ZMP is estimated by evaluating a performance limitation of ZMP feedback control, and the robot applies the predicted ZMP to detect imminent falls and to select correction motions. To further improve fall detection procedure Ogata et al. used discriminant analysis of experimental walking data labelled as fall and non-fall in order to construct a classifier [11].

Karssen et al. used multi-way principal component analysis (MPCA) in a simulation to predict the fall [12]. The method was able to predict whether the model is going to fall or not; in the case of a single disturbance, the method was able to predict the fall just within a single step after the disturbance. In addition, the method has an advantage of low implementation complexity and a low number of test runs.

Hobbelen et al. introduced Limit Cycle Walking paradigm for a bipedal locomotion with a Gait Sensitivity Norm as new disturbance rejection measure, which can be used as a robust fall indicator [13].

Kanoi and Hartland [14] investigated a use of Reservoir Computing for meta-sensor conception, involving generation of temporal meta-sensor for fall detection that was based on actual robot sensors. Their model was able to provide insight into robot status in the context of fall detections together with a low error rate. The model can be applied online and shown to be accurate in detecting instabilities that lead to robot falls.

Höhn and Gerth [15] proposed a probability-based balance monitoring concept with two algorithms that allows distinguishing between normal operation and instability. First algorithm uses Gaussian-Mixture Models (GMM) to describe the distribution of the robot's sensor data for two different states - stable locomotion or falling. Using this model and incoming sensory data it is possible to estimate the probability of the robot being in one of these states. The second algorithm is based on Hidden-Markov-Models (HMM), and the model is utilized in order to detect and identify unstable states using estimated parameters of their typical sequences in the robot's sensor data. Learning phase needed for estimation distribution densities and HMM parameters are generated with help of a simulation program. Robustness of the algorithms was tested in simulated experiments. The feature vectors of model were sampled every 10 ms within experiments. GMM and HMM algorithms took less than a millisecond on a desktop PC, that was also simulating the dynamical model of the robot. Hence, an online operation on the robot's microcontroller is feasible.

Goswami and Kalyanakrishnan introduced a system that uses supervised machine learning approach to achieve reliable fall prediction [20]. Learned solutions were compiled into decision lists within 16-dimensional feature space, and the false positive rate and lead-time tradeoff could be further controlled with internal parameters adjustment. Simulation of ASIMO-like robot were performed in order to verify the proposed solution.

Jeong-Jung Kim et al. proposed a state classification method for detecting falling with Support Vector machine (SVM) to classify the state [22]. SVM utilized sensor data from robot accelerometer and force sensing resistor (FSR) sensor. Training of the classifier was performed off-line and the trained classifier is used to classify the state of the biped robot in on-line mode. Robot simulator was used to verify the method. This approach was able to classify falling state within 0.01 s.

[illegible]fmann et al. proposed a fall protection system based on an artificial MLP neural network using a time series of gyroscope values [21]. Experiments were performed with low cost small-size robotic platform NAO, which unfortunately could not be immediately scaled up for human-size robots.

3 Fall Management

There are two primary objectives of dealing with the robot accidental fall over: (a) minimizing damage to the robot and (b) minimizing damage to objects in the vicinity of the fall. Strategy of reducing damage of the impact is primary when the robot is operating in a free space. On the contrary, in situations when the falling robot can cause injury to a person or damage to objects in its vicinity, the primary objective should be to eliminate such possibility.

While most researches treat fall as an unavoidable part of bipedal walking and focus on developing strategies to avoid falls and to minimize mechanical damage Wilken et al. [16] have investigated a deliberate fall of a humanoid soccer goalkeeper. Although their strategy to minimize fall damage consists mostly of mechanical solutions and concentrates on joints relaxation just before ground impact.

Another approach of fall damage minimizing utilizes heuristics such as manipulating the center of mass (CoM) of a robot. Ruiz-del Solar et al. [17] investigated several strategies, which were inherited from Japanese martial arts and are to be applied in the direction of a fall; each strategy concentrated on lowering the robot CoM. Each falling strategy produced a sequence of motions that modified the geometry of the robot body with intent to decrease the force of the impact, and spreading kinetic energy of the fall to transfer through a broader contact area. Based upon this research they implemented a low damage fall strategies for robots playing soccer [18].

Ishida et al. performed analysis of SDR-4X II robot fall, mimicking shock absorbers with servos using servo loop gain shift [19].

Fujiwara et al. in their comprehensive work with human-size HRP-2 robot, presenting their solutions for fall management in a series of publications [2–4]. They presented "UKEMI" strategy, a falling motion control that minimizes damage to a humanoid robot. This strategy employs optimal falling maneuvers to minimize impact forces and angular momentum. To minimize the landing impact of a falling motion, they use optimization technique based on variations calculation with a quadruple inverted pendulum model that was used to represent a falling motion. They tested the estimated optimal forward falling motion and obtained a smooth and damage-free fall [3]. Main drawback of the method is that it is based on an off-line optimization and does not support a real-time motion generation for humanoid robots in real environment. To conduct more experiments of falling over motion of HRP OpenHRP dynamics simulator was utilized [2]. With simulation, they obtained good estimates of robot states that are difficult to measure directly on a real robot, such as forces and moments acting on the hip joint, while robot does not have a six-axis force sensor at the hip. This knowledge is very important in order to design proper hip joint structure, which arguably is the most complex structure of a humanoid robot. Comparison with experimental results of a real humanoid robot demonstrated that an overall behavior of the robot simulated falling motion corresponds well with real experiments. Further experiments indicated that the impact force could be damped effectively even if the shock-absorbing features does not present in entire body of the robot. To decrease damages further, they have studied a balance between landing impact force and a position as well as the stability after landing and the position. Adding braking after the CoM lowering by extending the body just before the impact with the ground reduced the impact velocity. In order to make joints more compliant, the feedback gain reduction after braking was introduced [4].

Ogata et al. analyzed trajectories of CoM, both straight [11] and curvilinear [10]. Thus was performed for both phases of lowering the CoM and extending the body for reduction of the vertical impact speed.

Ruiz-del Solar et al., instead of minimizing the ground impact velocity, sought minimization of axial force and torque induced by the impact [17]. Using motion-capture data, a fall was analyzed and a human operator changed the joints positions to reduce the impact over the joints with maximal impulses.

In their study of intentional fall, Wilken et al. adopted an inverse approach. Instead of minimizing the fall damage, they first designed the fall motions and then changed the robot's structure to reduce the damage. Springs and flexible rubber struts were added to the most damage-prone locations of the robot was given the deliberate fall motion of a robot soccer goalkeeper [16].

Goswami et al. studied a control strategy for changing a default fall direction of a robot so that it could avoid collisions with surrounding people or objects in order to minimize damage to others. This strategy used the fact that the robot falling definitely happens at an edge of its support. The authors modified the position and orientation of this edge to change the fall trajectory to suit the environment. As the fall is predicted controller infers the optimal trajectory, which results in the safest fall. The fall controller was also enhanced with inertia shaping that changes robot's centroidal axis inertia [19].

Seung-kook Yun et al. proposed another approach to reduce damage to a humanoid robot during a fall [23]. Instead of finding an optimal configuration of the falling down robot, this strategy seeks to stop the robot from falling all the way to the ground, preventing full conversion of the robot's potential energy into kinetic energy, thus minimizing the force of impact. This is achieved via a sequence of three contacts with the ground of the swing foot and two hands. The final configuration resembles a tripod as it has a stable three-point contact with the ground with the robot's CoM above the ground.

Vincent Samy et al. proposed a fall strategy that combines two behaviors. The first behavior involves a closed-loop pose correction during the falling process, which would help to achieve best impact absorption. The second behavior performs a servo active compliance mode through instant PD gains reduction, instead of shutting-down or high-gains control. The authors suggested utilizing actuators as a spring-damper system by analyzing velocity, computing effective mass at the link's contact points and the motors characteristics [24].

Sehoon Ha et al. suggested another approach, which deals with planning of fall contact points in order to effectively dissipate momentum [25]. Given an unstable initial position, the planner searches for an optimal sequence of contact points such that the initial momentum is dissipated with minimal impacts. Rather than choosing from a collection of individual control strategies, the proposed method is a generic algorithm, which plans for appropriate maneuvers. Algorithm estimates the number of contacts, the order of contacts, and the position and time of contacts for supporting further momentum dissipation with a minimal damage to the robot.

A fall prevention system developed by Park et al. [30] used an inertial-measurement unit (IMU) to detect if the robot is falling or not. In the case of falling, the robot performed a forward step with a swing leg in order to prevent falling. Yet, the approach was tested in simulation experiments only.

4 Future Work: Fall Detection and Prevention for AR-601

The presented above algorithms are carefully tailored and verified in simulations and/or experimental work by their authors to support particular models of small-size or human-size robots. Practical implementation of each algorithms is not transferable to other models due to different hardware specifications and configuration, and, to the best of our knowledge, a generic solution does not exist yet. We avoid declaring an ambitious goal of suggesting such generic solution, but are interested to perform applied research on developing algorithms of fall detection and fall prevention that would maintain static and dynamic stability of our human-size robot AR-601.

Our target platform is bipedal robot AR-601 with 41 active degrees of freedom (DoF) that have been developed by Russian company "Android Technics" (Fig. 1, left). The total mass of the robot is 65 kg, the height is 1442 mm. Mass, and size parameters of the robot legs are given in Table 1, and for further hardware details about AR-601 the interested reader could refer to [31]. We had presented a virtual model of the robot in Matlab/Simulink environment together with a corresponding model in ROS/Gazebo environment (Fig. 1, right). These models were applied for modeling and algorithm evaluation, which utilized mass characteristics of the real robot, such as mass, CoM location and moments of inertia for each part. Locomotion control during the robot locomotion uses only 12 leg joints (6 DoF in each leg) driven by small electric motors with STM32F103T8U6 controllers and the communication protocol provides information about all motor states, pressure in robot's feet and on-board gyroscopes. Each leg consists of three joint axes in the hip, two joints in the ankle and one in the knee. We had modeled and experimentally verified dynamically stable AR-601 M robot locomotion with VHIPM and preview control methods [26].

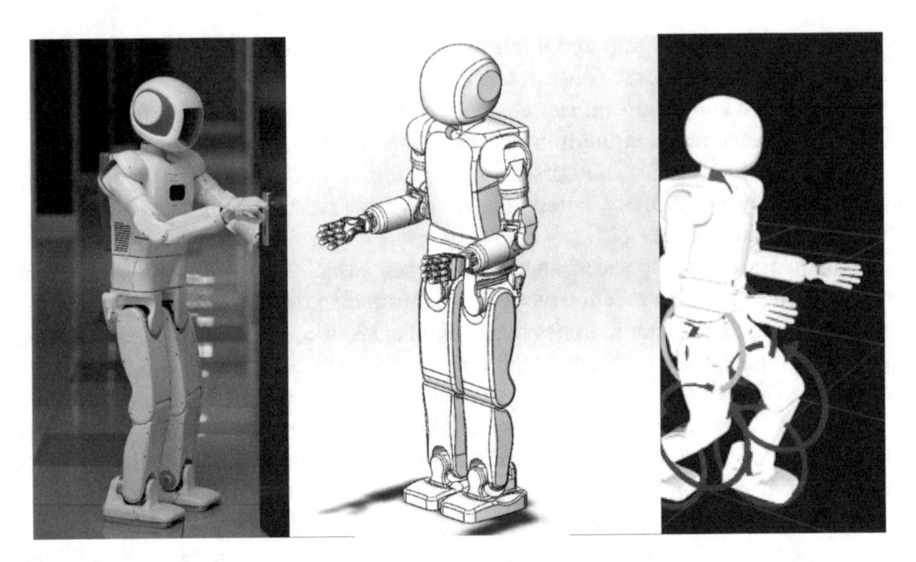

Fig. 1. Anthropomorphic robot AR-601 (left), its model in Solidworks (center) and ROS gazebo environment (right)

Table 1. Mass and size parameters of AR-601 legs

Link	Size parameters (mm)	Mass (kg)
Thigh	Length : 280	7.5
Shank	Length: 280	6.9
Foot	L × W × H: 254 × 160 × 106	3.2

During simulation in Matlab/Simulink and ROS/Gazebo environments that were followed by locomotion experiments, we encountered multiple instabilities that resulted in robot falls. These problems were persistent and required significant efforts during demonstrations in order to consistently preserve the robot, surrounding people and environment. To deal with this issue we acknowledge the acute need of diving into the field of fall detection and management. Using our model in virtual environment, we plan to test aforementioned algorithms to manage such situations.

5 Conclusions

In this paper we discussed different approaches to fall detection and management procedures for humanoid robots of small-size and human-size, which were verified in simulations and experimental work. Fall detection procedures are primarily based on various classification algorithms using supervised learning. For fall management, many approaches are centered on dissipating initial momentum of the fall using posture control. Another popular trend in fall management suggests using actuators to simulate shock absorbers.

We presented a brief overview of Russian bipedal robot AR-601, its modelling and simulation in Matlab/Simulink and ROS/Gazebo environments. Our future work concentrates on implementing fall detection and management procedures and verifying their performance for AR-601, both in intensive simulations and experiments.

Humanoid robot domain is not the only field, which deals with fall detection and management; these issues go far beyond robotics field and are important particularly in geriatric medicine. As a part of our long-term future work, we are interested to employ our insights on fall management in order to adapt humanoid robot algorithms for elderly support devices in order to improve their safety and quality of life.

Acknowledgments. Part of the work was performed according to the Russian Government Program of Competitive Growth of Kazan Federal University.

References

1. Goswami, A., Yun, S.K., Nagarajan, U., Lee, S.H., Yin, K., Kalyanakrishnan, S.: Direction-changing fall control of humanoid robots: theory and experiments. Auton. Robots **36**(3), 199–223 (2014)
2. Fujiwara, K., Kanehiro, F., Saito, H., Kajita, S., Harada, K., Hirukawa, H.: Falling motion control of a humanoid robot trained by virtual supplementary tests. In: IEEE International Conference on Robotics and Automation Proceedings, ICRA 2004, vol. 2, pp. 1077–1082, April 2004
3. Fujiwara, K., Kajita, S., Harada, K., Kaneko, K., Morisawa, M., Kanehiro, F., Hirukawa, H.: Towards an optimal falling motion for a humanoid robot. In: 2006 6th IEEE-RAS International Conference on Humanoid Robots, pp. 524–529, February 2007
4. Fujiwara, K., Kajita, S., Harada, K., Kaneko, K., Morisawa, M., Kanehiro, F., Hirukawa, H:. An optimal planning of falling motions of a humanoid robot. In: IEEE/RSJ International Conference on Intelligent Robots and Systems, IROS 2007, pp. 456–462, October 2007
5. Chia, P.C., Lee, C.H., Chen, T.S., Kuo, C.H., Lee, M.Y., Chen, C.M.S.: Correlations of falling signals between biped robots and humans with 3-axis accelerometers. In: International Conference on System Science and Engineering (ICSSE), pp. 509–514, June 2011
6. Forner Cordero, A.: Human Gait, Stumble and... fall? Mechanical limitations of the recovery from a stumble. Universiteit Twente (2003)
7. Cordero, A.F., Koopman, H.F.J.M., Van der Helm, F.C.T.: Multiple-step strategies to recover from stumbling perturbations. Gait Posture **18**(1), 47–59 (2003)
8. Cordero, A.F., Koopman, H.J.F.M., van der Helm, F.C.: Mechanical model of the recovery from stumbling. Biol. Cybern. **91**(4), 212–220 (2004)
9. Renner, R., Behnke, S.: Instability detection and fall avoidance for a humanoid using attitude sensors and reflexes. In: IEEE/RSJ International Conference on Intelligent Robots and Systems, pp. 2967–2973, October 2006
10. Ogata, K., Terada, K., Kuniyoshi, Y.: Real-time selection and generation of fall damage reduction actions for humanoid robots. In: 8th IEEE-RAS International Conference on Humanoid Robots, Humanoids, pp. 233–238, December 2008

11. Ogata, K., Terada, K., Kuniyoshi, Y.: Falling motion control for humanoid robots while walking. In: 7th IEEE-RAS International Conference on Humanoid Robots, pp. 306–311, November 2007
12. Karssen, J.D., Wisse, M.: Fall detection in walking robots by multi-way principal component analysis. Robotica **27**(02), 249–257 (2009)
13. Hobbelen, D.G., Wisse, M.: A disturbance rejection measure for limit cycle walkers: the gait sensitivity norm. IEEE Trans. Robotics **23**(6), 1213–1224 (2007)
14. Kanoi, R., Hartland, C.: Fall detections in humanoid walk patterns using reservoir computing based control architecture. In: 5th National Conference on Control Architecture of Robots, May 2010
15. Höhn, O., Gerth, W.: Probabilistic balance monitoring for bipedal robots. Int. J. Robotics Res. **28**(2), 245–256 (2009)
16. Wilken, T., Missura, M., Behnke, S.: Designing falling motions for a humanoid soccer goalie. In: International Conference on Humanoid Robots Proceedings of the 4th Workshop on Humanoid Soccer Robots (2009)
17. Ruiz-del-Solar, J., Moya, J., Parra-Tsunekawa, I.: Fall detection and management in biped humanoid robots. In: 2010 IEEE International Conference on Robotics and Automation (ICRA), pp. 3323–3328, May 2010
18. Ruiz-del-Solar, J., Palma-Amestoy, R., Marchant, R., Parra-Tsunekawa, I., Zegers, P.: Learning to fall: designing low damage fall sequences for humanoid soccer robots. Robotics Auton. Syst. **57**(8), 796–807 (2009)
19. Nagarajan, U., Goswami, A.: Generalized direction changing fall control of humanoid robots among multiple objects. In: 2010 IEEE International Conference on Robotics and Automation (ICRA), pp. 3316–3322, May 2010
20. Kalyanakrishnan, S., Goswami, A.: Learning to predict humanoid fall. Int. J. Humanoid Robotics **8**(02), 245–273 (2011)
21. Hofmann, M., Schwarz, I., Urbann, O., Ziegler, F.: A Fall Prediction System for Humanoid Robots Using a Multi-Layer Perceptron
22. Kim, J.J., Choi, T.Y., Lee, J.J.: Falling avoidance of biped robot using state classification. In: IEEE International Conference on Mechatronics and Automation, ICMA 2008, pp. 72–76, August 2008
23. Yun, S.K., Goswami, A.: Tripod fall: concept and experiments of a novel approach to humanoid robot fall damage reduction. In: 2014 IEEE International Conference on Robotics and Automation (ICRA), pp. 2799–2805, May 2014
24. Samy, V., Kheddar, A.: Falls control using posture reshaping and active compliance. In: 2015 IEEE-RAS 15th International Conference on Humanoid Robots (Humanoids), pp. 908–913, November 2015
25. Ha, S., Liu, C.K.: Multiple contact planning for minimizing damage of humanoid falls. In: 2015 IEEE/RSJ International Conference on Intelligent Robots and Systems (IROS), pp. 2761–2767, September 2015
26. Khusainov, R., Afanasyev, I., Sabirova, L., Magid, E.: Bipedal robot locomotion modelling with virtual height inverted pendulum and preview control approaches in Simulink environment. J. Robotics Netw. Artif. Life **3**(3), 182–187 (2016)
27. Wright, J., Jordanov, I.: Intelligent approaches in locomotion. In: The 2012 International Joint Conference on Neural Networks (IJCNN), pp. 1–8, June 2012
28. Hurmuzlu, Y., Génot, F., Brogliato, B.: Modeling, stability and control of biped robots—a general framework. Automatica **40**(10), 1647–1664 (2004)
29. Asano, F., Luo, Z.W.: Energy-efficient and high-speed dynamic biped locomotion based on principle of parametric excitation. IEEE Trans. Robotics **24**(6), 1289–1301 (2008)

30. Park, G.-M., Baek, S.-H., Kim, J.-H.: Falling prevention system from external disturbances for humanoid robots. In: Kim, J.-H., Yang, W., Jo, J., Sincak, P., Myung, H. (eds.) Robot Intelligence Technology and Applications 3. AISC, vol. 345, pp. 97–105. Springer, Cham (2015). doi:10.1007/978-3-319-16841-8_10
31. Khusainov, R., Shimchik, I., Afanasyev, I., Magid, E.: 3D modelling of biped robot locomotion with walking primitives approach in Simulink environment. In: Filipe, J., Madani, K., Gusikhin, O., Sasiadek, J. (eds.) Informatics in Control, Automation and Robotics 12th International Conference, ICINCO 2015 Colmar, France, July 21-23, 2015 Revised Selected Papers. LNEE, vol. 383, pp. 287–304. Springer, Cham (2016). doi:10.1007/978-3-319-31898-1_16

Business Process Management

Modelling of the Logistic Supplier-Consumer Behavior

Petr Suchánek[1(✉)] and Robert Bucki[2]

[1] Silesian University in Opava, School of Business Administration in Karviná, Karviná, Czech Republic
suchanek@opf.slu.cz
[2] Institute of Management and Information Technology, Bielsko-Biala, Poland
rbucki@wsi.net.pl

Abstract. The paper highlights the problems of mathematical modelling in the delivery system. The system describes the suppliers who offer different types of products as well as the consumers who order different products. Products are ordered at stochastic times, however, manufacturers offer predictable demand. The problem becomes more complex when the number of orders grows. The structure of the system is shown, equations of state are introduced and control algorithms as well as criteria are proposed. Orders change their state which leads to modifying it at every decision stage. The same concerns the actual output of manufacturers which also has to be modified. Therefore, the problem consists of the design of such a delivery pattern which can minimise losses of the discussed company. The goal of the paper is to present the mathematical model of the logistic system taking into account the consumer-supplier relations. The model forms the basis for the subsequent information support tool.

Keywords: Logistic modelling · Mathematical modelling · Delivery system · Information support · Computational modelling · Simulation · Heuristic algorithms · Business process management · Optimisation

1 Introduction

One of the important parts of the value and logistic chain is the method of delivery of products from the manufacturer to the customer. Each delivery process must guarantee customer's satisfaction on one side and the simplest method of delivery for suppliers on the other. The second of these conditions is very important because it is a significant factor affecting the overall price of the product. The primary assumption is the correct definition of logistics and general business processes as well as management and measurement of their effectiveness [1]. The effectiveness of the supply chain is in most cases assessed on the basis of time and price. Both of these parameters are influenced by many other factors, which may be a number of interim storage on the way from the producer to the end consumer, the number of employees, the number of delivered products and the demands on their transport providers (there is a difference between transporting a heavy industry product, for example a ship propeller, and conventional goods sold to customers for example within the e-commerce [2]), environmental

G. Jezic et al. (eds.), *Agent and Multi-Agent Systems: Technology and Applications*, Smart Innovation, Systems and Technologies 74,
DOI 10.1007/978-3-319-59394-4_21

requirements [3], the necessity to apply reverse logistics [4] etc. Thus, an important aspect is economising and optimising logistic processes [5, 6], which ultimately contributes to the satisfaction of customers' psychological needs who can get products at a lower price [7]. Nowadays, process optimisation is almost always carried out using computer simulations for which the starting points are mathematical models. Models can be created on the basis of many approaches and methods. One of them is a heuristic approach which, despite some shortcomings presented for example in [8], can be used in many cases and can help the company find a better structure and the functionality of the company's system of processes [9–13]. In the area of process optimisation initially we start from the map of business processes and principles of business process management [14], for which mathematical simulation models are created. Based on simulation results, we can carry out process optimisation, so in this case, optimisation of logistic processes [15]. In connection with the foregoing, the aim of the article is to present a mathematical simulation model which takes into account different ways of supplying goods to customers, state equations of the model, control algorithms of the delivery process and decision criteria. The background of the model is formed by taking into account the estimated amount of financial implementation costs i.e. the factor affecting the final price of the product.

2 Mathematical Model

To discuss the problem of delivering products to customers we have to analyse the following actions which form the basis of our reasoning:

i. delivering products directly to customers;
ii. delivering products to the storing space of the company;
iii. decommissioning products;
iv. storing products for a specified period of time;
v. commissioning products;
vi. delivering products to customers from the storing space of the company.

Each of the above mentioned activities generates costs. These costs should be minimised in order to let the company offer products at the competitive price. Products can be delivered to customers directly and it is expected that this kind of activity may reduce the costs of servicing the order, however, if more products are to be delivered to one or more customers, it can pose a problem which can be solved by means of the intermodal logistic transport centre in the form of the company storage space. So the following situation can be analysed:

i. products of the same type produced by one manufacturer are delivered to one customer;
ii. products of the same type produced by one manufacturer are delivered to multiple customers;
iii. more products of various types produced by one manufacturer are delivered to one customer;

iv. more products of various types produced by one manufacturer are delivered to multiple customers.

Let us define the structure matrix of the outgoing supply chain (1):

$$E_{out}^{k} = \left[e_{m,n}^{k}\right],\ m = 1,\ldots,M,\ n = 1,\ldots,N,\ k = 0,1,\ldots,K \tag{1}$$

$e_{m,n}^{k}$ - the information whether or not the n-th order of the m-th customer at the k-th stage is valid. The elements of the structure matrix of the outgoing supply chain take the following values:$e_{m,n}^{k} = 1$ if the n-th order of the m-th customer at the k-th stage exists; $e_{m,n}^{k} = 0$ otherwise.

The structure of the system (1), the outgoing supply chain (3), the incoming supply chain (4), the state of order (25), the storage state of the system (21) as well as the storage capacity (23) are described at the k-th stage, $k = 0, 1, \ldots, K$. If there is any activity in the system, the k-th stage is subject to modification.

Let us define the structure matrix of the incoming supply chain (2):

$$E_{in}^{k} = \left[e_{i,n}^{k}\right],\ i = 1,\ldots,M,\ n = 1,\ldots,N,\ k = 0,1,\ldots,K \tag{2}$$

$e_{i,n}^{k}$ - the information whether or not the n-th ready product of the i-th producer at the k-th stage is available. The elements of the structure matrix of the incoming supply chain take the following values: $e_{i,n}^{k} = 1$ if the n-th ready product of the i-th producer at the k-th stage exists; $e_{i,n}^{k} = 0$ otherwise.

Let us define the order matrix of the outgoing supply chain (3):

$$Z_{out}^{k} = \left[z_{m,n}^{k}\right],\ m = 1,\ldots,M,\ n = 1,\ldots,N,\ k = 0,1,\ldots,K \tag{3}$$

$z_{m,n}^{k}$ - the number of the n-th order elements of the m-th customer at the k-th stage.

Let us define the matrix of products of the incoming supply chain (4):

$$Z_{in}^{k} = \left[z_{i,n}^{k}\right],\ i = 1,\ldots,I,\ n = 1,\ldots,N,\ k = 0,1,\ldots,K \tag{4}$$

$z_{i,n}^{k}$ - the number of the n-th ready product elements available from the i-th producer at the k-th stage.

The following assumptions must be made:

(1) No matter which producer manufactures product of the n-th type they are stored at the same storage place.
(2) The n-th product elements can be accepted by the company only on condition that their number does not exceed the allowable capacity.
(3) Products are stored in accordance with either the FIFO or LIFO method.

Let us define the matrix of transfer possibilities from the producer to the customer (avoiding the storage space of the discussed company) (5):

$$\Omega_{i,m} = \left[\omega_{i,m}\right],\ i = 1,\ldots,I,\ m = 1,\ldots,M \tag{5}$$

where: $\omega_{i,m}$ - the adjustment of the i-th producer to the m-th customer. At the same time: $\omega_{i,m} = 1$ if the adjustment is possible, otherwise $\omega_{i,m} = -1$ and the i-th producer products have to be stored in the company storage space in order to be sold to the m-th customer.

Let us define the matrix of storing possibilities in the company marshalling store (6):

$$\Omega_n = \left[\omega_n\right],\ n = 1,\ldots,N \tag{6}$$

where: ω_n - storing possibilities of the n-th product.

At the same time: $\omega_n = 1$ if storing the n-th product is possible, otherwise $\omega_n = -1$ and the n-th product has to be transported directly to the m-th customer from the i-th producer.

Let us introduce the matrix of transport times of products to the customers avoiding the storage space of the company (7):

$$T_{i,n\to m,n} = \left[\tau_{i,n\to m,n}\right],\ i = 1,\ldots,I,\ m = 1,\ldots,M\,,n = 1,\ldots,N \tag{7}$$

where: $\tau_{i,n\to m,n}$ - the transport time of the n-th product of the i-th producer to the m-th customer.

Let us introduce the matrix of transport times of products to the storage space of the company (loading and unloading times are included) (8):

$$T_{i,n\to\Delta} = \left[\tau_{i,n\to\Delta}\right],\ i = 1,\ldots,I,\ n = 1,\ldots,N \tag{8}$$

where: $\tau_{i,n\to\Delta}$ - the transport time of the n-th product of the i-th producer to the storage space of the company.

Let us introduce the matrix of transport times of products from the company storage space to dedicated customers (loading and unloading times are included) (9):

$$T_{\Delta\to m,n} = \left[\tau_{\Delta\to m,n}\right],\ m = 1,\ldots,M,\ n = 1,\ldots,N \tag{9}$$

where: $\tau_{\Delta\to m,n}$ - the transport time of the n-th product to the m-th customer.

Let us introduce the matrix of average transport costs of products from producers to the customer's storehouse (avoiding the storage space of the company) (10):

$$C_{i,n\to m,n} = \left[c_{i,n\to m,n}\right],\ i = 1,\ldots,I,\ m = 1,\ldots,M,\ n = 1,\ldots,N \tag{10}$$

where: $c_{i,n\to m,n}$ - the transport cost of the n-th product of the i-th producer to the m-th customer.

Let us introduce the matrix of average transport costs of products from producers to the company storage space (11):

$$C_{i,n\to\Delta} = \left[c_{i,n\to\Delta}\right],\ i = 1,\ldots,I,\ n = 1,\ldots,N \tag{11}$$

where: $c_{i,n\to\Delta}$ - the transport cost of the n-th product of the i-th producer to the company storage space.

Let us introduce the matrix of average transport costs of products from the company to customers:

$$C_{\Delta\to m,n} = \left[c_{\Delta\to m,n}\right],\ i = 1,\ldots,I,\ n = 1,\ldots,N \tag{12}$$

where: $c_{\Delta\to m,n}$ - the transport cost of the n-th product to the m-th customer.

Storing products generates costs depending on the time of storing which are shown in the vector below (13):

$$C_{st} = \left[\varphi \cdot c_{st\_n}\right],\ n = 1,\ldots,N \tag{13}$$

where: c_{st_n} - the unitary cost of storing the n-th product,

φ - the number of time units during which the n-th product is stored in its dedicated storing place.

For illustration purposes the problem is analysed from the point of view of the load capacity expressed in units. Therefore, the maximal load vector of means of transport is introduced (14):

$$H = \left[h_u\right],\ u = 1,\ldots,U \tag{14}$$

where: h_u - the maximal load of the u-th means of transport expressed in units.

The capacity matrix of current load of means of transport (15):

$$B = \left[b^k_{n,u}\right],\ k = 0,1,\ldots,K,\ n = 1,\ldots,N,\ u = 1,\ldots,U \tag{15}$$

where: $b^k_{n,u}$ - the current load of the u-th means of transport with the n-th type product at the k-th stage.

At the same time $\sum_{n=1}^{N} v_{n,u} \le h_u,\ n = 1,\ldots,N,\ u = 1,\ldots,U$

where: $v_{n,u}$ - the number of the n-th product units which can be loaded on the u-th means of transport.

It is assumed that if more products of the same n-th type are to be dispatched by the same means of transport their transport time equals the transport time of only one n-th type product on condition that the load of the means of transport is not exceeded.

3 Decommissioning and Commissioning

Let us introduce the matrix of decommissioning times of products to the storage space of the company (16):

$$T_{dec\_n} = [\tau_{dec\_n}],\ n = 1, \ldots, N \tag{16}$$

where: τ_{dec_n} - the decommissioning time of the n-th product including transporting it to its storage place.

Let us introduce the matrix of commissioning times of products in the storage space of the company (17):

$$T_{com\_n} = [\tau_{com\_n}],\ n = 1, \ldots, N \tag{17}$$

where: τ_{com_n} - the commissioning time of the n-th product including transporting it to dispatch place.

Let us introduce the vector of average decommissioning costs of products from producers (18):

$$C_{dec\_n} = [c_{dec\_n}],\ n = 1, \ldots, N \tag{18}$$

where: c_{dec_n} - the decommissioning cost of the n-th product in the company storage space.

Let us introduce the vector of average commissioning costs of products before dispatching them to customers (19):

$$C_{com\_n} = [c_{com\_n}],\ n = 1, \ldots, N \tag{19}$$

where: c_{com_n} - the commissioning cost of the n-th product.

Matrices of delivery costs depending on different combinations of producers, products and customers are included in Table 1.

Table 1. Delivery costs

Delivery costs	Mathematical representation
For multiple orders directly from various producers to the m-th customer	$C_{I,N\to m,N} = \sum_{n=1}^{N}\sum_{i=1}^{I} c_{i,n\to m,n}$
For various products of the same n-th producer directly to multiple customers	$C_{I,n\to M,n} = \sum_{m=1}^{M}\sum_{i=1}^{I} c_{i,n\to m,n}$
For various products of various producers directly to multiple customers	$C_{I,N\to M,N} = \sum_{m=1}^{M}\sum_{n=1}^{N}\sum_{i=1}^{I} c_{i,n\to m,n}$
For various orders directly from various producers to multiple customers via the company storage space	$C_{\Delta} = \sum_{m=1}^{M}\sum_{n=1}^{N}\sum_{i=1}^{I} (c_{i,n\to\Delta} + c_{dec\_n} + c_{st\_n} + c_{com\_n} + c_{\Delta\to m,n})$

4 Equations of State

Let us define the vector of storage capacity of the company (20):

$$G = [g_n],\ n = 1, \ldots, N \tag{20}$$

where: g_n^k - the maximal number of the n-th product elements which can be stored in the marshalling store of the company.

The vector of storage state of the company takes the following form (21):

$$S^k = \left[s_n^k\right],\ k = 0, 1, \ldots, K,\ n = 1, \ldots, N \tag{21}$$

where: s_n^k - the number of the n-th type products which are already stored in the marshalling store of the company at the k-th stage.

The vector of storage capacity of the company takes the following form (22):

$$P^k = \left[p_n^k\right],\ k = 0, 1, \ldots, K,\ n = 1, \ldots, N \tag{22}$$

where: p_n^k - the number of the n-th type products which still can be stored in the marshalling store of the company at the k-th stage.

The storage capacity is calculated on basis of the following Eq. (23):

$$p_n^k = g_n - s_n^k \tag{23}$$

The state of the order matrix of the outgoing supply chain changes after every decision as follows (24):

$$\begin{array}{ccccccccccc} Z_{out}^0 & \rightarrow & Z_{out}^1 & \rightarrow & \cdots & \rightarrow & Z_{out}^k & \rightarrow & \cdots & \rightarrow & Z_{out}^K \\ & \uparrow & & \uparrow & & \uparrow & & \uparrow & & \uparrow & \\ & \Delta Z_{out}^0 & & \Delta Z_{out}^1 & & \Delta Z_{out}^{k-1} & & \Delta Z_{out}^k & & \Delta Z_{out}^{K-1} & \end{array} \tag{24}$$

The state of the order matrix element of the outgoing supply chain does not change when:$z_{m,n}^k = z_{m,n}^{k-1}$ if the m-th customer does not require the n-th product at the k-th stage; $z_{m,n}^k \neq z_{m,n}^{k-1}$ otherwise. The order matrix elements are subject to modification at each stage.

The state of the order matrix of the incoming supply chain changes after every decision as follows (25):

$$\begin{array}{ccccccccccc} Z_{in}^0 & \rightarrow & Z_{in}^1 & \rightarrow & \cdots & \rightarrow & Z_{in}^k & \rightarrow & \cdots & \rightarrow & Z_{in}^K \\ & \uparrow & & \uparrow & & \uparrow & & \uparrow & & \uparrow & \\ & \Delta Z_{in}^0 & & \Delta Z_{in}^1 & & \Delta Z_{in}^{k-1} & & \Delta Z_{in}^k & & \Delta Z_{in}^{K-1} & \end{array} \tag{25}$$

At the same time the state of the order matrix element of the incoming supply chain does not change when:$z_{i,n}^k = z_{i,n}^{k-1}$ if the i-th producer does not supply the n-th product at the k-th stage; $z_{i,n}^k \neq z_{i,n}^{k-1}$ otherwise.

5 Control Algorithms

There are usually more orders at the same time so the dispatcher has to decide which order can be send first to the chosen customer. To meet this requirement there is a need to introduce algorithms responsible for advising the operator how to carry out the dispatching process at various stages of delivery (see Table 2).

Table 2. Delivery algorithms

Delivery algorithms	Description
$\alpha_c = \min c_{i,n \to m,n}$	- adjusts the i-th producer's n-th product to the m-th customer characterised by the minimal delivery costs; - avoids the company storage space
$\alpha_\tau = \min \tau_{i,n \to m,n}$	- adjusts the i-th producer's n-th product to the m-th customer characterised by the minimal delivery time; - avoids the company storage space
$\alpha_{c_{I,N \to m,N}} = \min c_{i,n \to m,n}$	- adjusts producers and their products to the m-th customer characterised by the minimal delivery costs; - avoids the company storage space
$\alpha_{c_{I,n \to M,n}} = \min c_{i,n \to m,n}$	- adjusts products of the n-th producer to various customers characterised by the minimal delivery costs; - avoids the company storage space
$\alpha_{c_{I,N \to M,N}} = \min c_{i,n \to m,n}$	- adjusts products of all producers to various customers characterised by the minimal delivery costs; - avoids the company storage space
$\alpha_{c_\Delta} = \min \Delta$	- adjusts products of all producers to various customers characterised by the minimal delivery costs; - uses the company storage space; $\Delta = c_{i,n \to \Delta} + c_{dec\_n} + c_{st\_n} + c_{com\_n} + c_{\Delta \to m,n}$

6 Criteria

Criteria take into account delivery time, costs and load (see Tables 1 and 2).

There are the following criteria proposed:

1. The delivery time criterion: $Q_\tau \to \min$

   (i.e. the sum of all separate delivery times in the system tends to be minimal).

2. The delivery cost criterion: $Q_c \to \min$

   (i.e. the sum of all separate delivery costs in the system tends to be minimal).

3. The load criterion: $Q_{load} \to \max$

   (i.e. the sum of all separate loads throughout the transport process in the system tends to be maximal).

7 Conclusions

The paper is devoted to mathematical modelling of the highly complex economic system placed in the supply chain. The system can be used for representing a delivery company problem either with or without its own storage space. The structure matrices can be adjusted according to the actual needs of the company. Customers' orders and manufacturers' products are verified at each stage. The capacity of the storage space is limited, so it is often necessary to avoid storing products in it. Instead, it is convenient to transport products directly to customers which can increase profits of the delivery company. The problem with modelling such a system seems to be more complicated as previously thought as customers may have differentiated priorities. The model is the initial work which is expected to be developed in further works. Moreover, the simulator of the system should lead to approximations letting us achieve practical assumptions. Simulating large initial data could mean verifying the correctness of the proposed heuristic algorithms.

Acknowledgement. This paper was supported by the project SGS/19/2016 at the Silesian University in Opava – Advanced Mining Methods and Simulation Techniques in the Business Process Domain.

References

1. Koc, T., Bozdag, E.: Measuring the degree of novelty of innovation based on Porter's value chain approach. Eur. J. Oper. Res. **257**(2), 559–567 (2017)
2. Masmoudi, M., Benaissa, M., Chabchoub, H.: Mathematical modelling for a rich vehicle routing problem in e-commerce logistics distribution. In: International Conference on Advanced Logistics and Transport (ICALT), Sousse, pp. 290–295 (2013)
3. Korczak, J., Kijewska, K., Iwan, S.: Strategic aspects of an eco-logistic chain optimization. Sustainability **8**(4), 277–287 (2016)
4. Jurova, M., Jurica, P.: Management of reverse logistics in business processes. In: Innovation Management and Sustainable Economic Competitive Advantage: From Regional Development to Global Growth, Madrid, pp. 2266–2271 (2015)
5. Bucki, R., Chramcov, B.: Economizing logistic costs of the manufacturing system using mathematical modelling to aid decision-making. In: 33rd International Conference Mathematical Methods in Economics (MME 2015), Cheb, pp. 80–85 (2015)
6. Goldsby, T.J., Zinn, W.: Technology innovation and new business models: can logistics and supply chain research accelerate the evolution? J. Bus. Logistics **37**(2), 80–81 (2016)
7. Wilson, C.M., Price, C.W.: Do consumers switch to the best supplier? Oxford Econ. Papers-New Ser. **62**(4), 647–668 (2010)
8. Zhang, X.X., Huang, B., Tay, R.: Estimating spatial logistic model: a deterministic approach or a heuristic approach? Inf. Sci. **330**(SI), 358–369 (2016)
9. Gonzalez-Ramirez, R.G., Smith, N.R., Askin, R.G., Kalashinkov, V.: A heuristic approach for a logistics districting problem. Int. J. Innovative Comput. Inf. Control **6**(8), 3551–3562 (2010)
10. Melo, M.T., Nickel, S., Saldanha-da-Gama, F.: An efficient heuristic approach for a multi-period logistics network redesign problem. TOP **22**(1), 80–108 (2014)

11. Bucki, R., Chramcov, B., Suchanek, P.: Heuristic algorithms for manufacturing and replacement strategies of the production system. J. Univers. Comput. Sci. **21**(4), 503–525 (2015)
12. Bucki, R., Suchanek, P.: Information support for solving the order priority problem in business logistic systems. In: 34th International Conference Mathematical Methods in Economics (MME 2016), pp. 85–90. Tech UnivLiberec, Fac Econ, Fac Mech Engn, Liberec (2016)
13. Suchánek, P., Bucki, R.: Business process modeling of logistic production systems. In: Jezic, G., Chen-Burger, Y.-H.J., Howlett, Robert J., Jain, Lakhmi C. (eds.) Agent and Multi-Agent Systems: Technology and Applications. SIST, vol. 58, pp. 199–207. Springer, Cham (2016). doi:10.1007/978-3-319-39883-9_16
14. Bozic, D., Stankovic, R., Rogic, K.: Possibility of applying business process management methodology in logistic processes optimization. Promet-Traffic & Transp. **26**(6), 507–516 (2014)
15. Jasek, R., Sedlacek, M., Chramcov, B., Dvorak, J.: Application of simulation models for the optimization of business processes. In: Proceedings of the International Conference on Numerical Analysis and Applied Mathematics 2015 (ICNAAM 2015), Rhodes, pp. 120028-1–120028-4 (2015)

Conversion of Real Data from Production Process of Automotive Company for Process Mining Analysis

Miroslav Dišek, Roman Šperka(✉), and Jan Kolesár

Department of Business Economics and Management,
School of Business Administration in Karviná, Silesian University in Opava,
Univerzitní nám. 1934/3, 733 40 Karviná, Czech Republic
{0150611, sperka, 0160358}@opf.slu.cz

Abstract. The aim of this paper is to convert the real data from the raw format from different information systems (log files) to the format, which is suitable for process mining analysis of a production process in a large automotive company. The conversion process will start with the import from several relational databases. The motivation is to use the DISCO tool for importing real pre-processed data and to conduct process mining analysis of a production process. DISCO generates process models from imported data in a comprehensive graphical form and provides different statistical features to analyse the process. This makes it possible to examine the production process in detail, identify bottlenecks, and streamline the process. The paper firstly presents a brief introduction of a manufacturing process in a company. Secondly, it provides a description of a conversion and pre-processing of chosen real data structures for the DISCO import. Then, it briefly describes the DISCO tool and proper format of pre-processed log file, which serves as desired input data. This data will be the main source for all consecutive operations in generated process map. Finally, it provides a sample analysis description with emphasis on one production process (process map and few statistics). To conclude, the results obtained show high demands on pre-processing of real data for suitable import format into DISCO tool and vital possibilities of process mining methods to optimize a production process in an automotive company.

Keywords: Process mining · Data cleaning · Data cleaning tools · DISCO

1 Introduction

The aim of this paper is to describe the real data pre-processing stage into suitable import format, which will be used for process mining analysis in a DISCO software tool. The subject of the data is a real production process in a Czech automotive company.

Process mining aims at a detection, analysis and optimization of business processes based on the data from log files, available in companies' information systems (IS). Process mining represents the missing link between traditional business process analysis and data mining [1]. Pre-processing of real data is an important task [2], which main purpose is to ensure correct data input for a learning or analysis phase of process

G. Jezic et al. (eds.), *Agent and Multi-Agent Systems: Technology and Applications*, Smart Innovation, Systems and Technologies 74,
DOI 10.1007/978-3-319-59394-4_22

mining. Its main role is to determine, which target attributes and information in a business process are essential, and which information could be omitted, while not corrupting the nature of a business process [3]. Note, that the person, which is familiar with the nature of the business process and analysed domain, plays an important role in the pre-processing activity.

Poor quality of pre-processed data is what causes losses. For example, once we have in the system post address, to which the mail carrier can´t deliver the goods, it is a poor data entry. Once we have an incorrect identification number of the customer, but the customer can be identified by his customer number in the system, we do not consider it to be a poor data entry. The quality of the data is determined by its use. Data simply can´t be judged just by themselves without the context [4].

Tools for automated data cleaning [5] have been existing for years. They usually focus on a specific database areas, which define possible values that can be inserted in each field or attribute, such as name and address entry fields. Typically, they use a set of matching rules from a library or the user delivers the rules interactively. Their task is to verify street names, city names and zip codes and they transform existing data into individual standard elements. They use record unification to determine whether two data records are about the same topic and are able to combine the individual records, which are connected with, e.g., the same address. Cleaning data tools [6] can be different in the level of sophistication of audit data, cleaning and migration.

This paper represents a partial research within the project of Silesian University in Opava, Czech Republic "Advanced mining methods and simulation techniques in business process domain." This project initiates partial interdisciplinary research in the field of simulation of business processes and process mining, which takes place at the Department of Mathematics and Informatics, and Department of Business economics and Management with the goal to engage doctoral students from Business economics and Management specialization, and students of the master's program Managerial informatics in scientific research. In cooperation with enterprises from Czech Republic, real data about business processes that are stored in the information systems log files will be analysed. In this paper, we use specific real data from a large automotive company, which business is a production of trucks and we focus on selected parts of a production process in the company, divided into several sub-processes. These processes generate data, which are gradually stored in relational databases and our task was to convert the data from these relational databases and rationalize them to conduct process mining analysis. The aim of the analysis is to find possible dependencies of activities performed in the production process of the company and to suggest process optimization or re-design. We will use graphical process model based on control-flow and statistics provided by DISCO to ensure the optimization by identifying weak places in the production process.

The structure of the paper is as follows. First section deals with familiarization with the entire manufacturing process and data collection of the company. Next, we introduced the structure of data stored in a database on SQL Server. Final sections describe the data conversion and pre-processing for process mining. Based on this phase, the production process data will be imported into DISCO software tool and consequently it will be briefly analysed.

2 Manufacturing Process

The motto of the automotive company, we are dealing with is: "Everything is related to everything". The company belongs to the world known manufacturers operating in automotive industry with a truck specialization. Its main concern are not only casual customers, but they are also producing custom-made vehicles, and vehicles for military needs. By the end of 2015, the company was operating with approximately 850 own employees and many others from personal agencies and branches. The volume of production in 2015 was 821 complete vehicles. The manufacturing process of the company is further described in the following text. It consists of several production processes executed in different production halls. They are using the name "economic centre" instead of production hall in the company. The entire production takes place in four economic centres with different production processes. We will further use real data from one production process dealing with transmission parts in the last part of this paper (Figure 1).

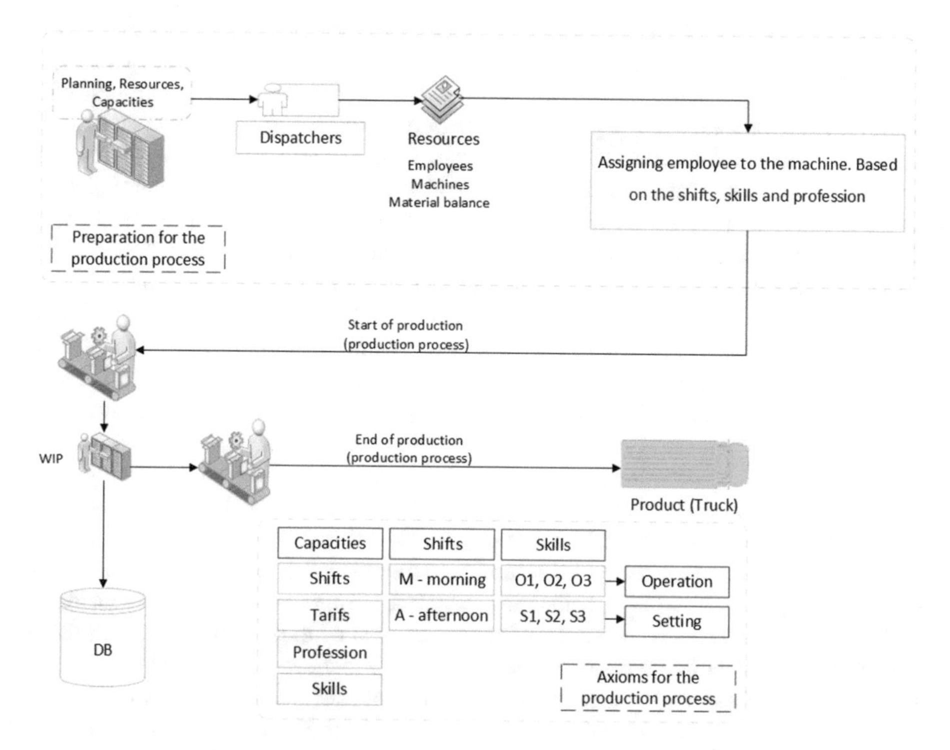

Fig. 1. Manufacturing process workflow. (Source: own)

The entire manufacturing process is dependent on the material balance, its flow, and production planning. Production planning is mainly influenced by production orders coming from customers, inclusive their deadlines. They also use a forecasting activity to control the production volume for planned time period. Production process is controlled by operative workers, which use several information systems to plan the production. They take necessary interventions when real manufacturing process differentiates from desired values. There are two main indicators, which are necessity to control and the entire manufacturing process is based on, namely, material and manufacturing order.

The manufacturing process is divided into following parts, which constitute the overall workflow:

- Planning and forecasting of production
- Operative – production dispatching
- Main resources – human work, machines
- Material flow and balance
- Production process – production cycles
- The creation of data entries and communication with production IS

The workflow of the process is composed from chronological operations/activities and many consecutive conditions, which are dependent to each other and perfectly fulfil conditions set. Right before the production starts and operative workers start planning, they have to align all operations and understand that in 96% of cases one machine is able to manage exactly one operation. The rest are composed with multi-machine operations. The production operative and production planner starts with aligning all production orders data according to their finishing date. After aligning all data the planner also checks the production orders, which although have their finishing data too but their preparation or material is most of the time atypical. Next, all production orders are divided based on their type, economic centre and chief production team. The following step is to choose appropriate machines, which have their specific operation rules. Employees, which are able to fulfil service conditions of chosen machine and the time window for working with them, continue to the shorter list of candidates for according machine. Note that production hall planners have to assign employees in a way to achieve the highest fluency of work continuously. Because of a huge size of the economic centre, it is necessary to properly place qualified workers in order they don't have to move across the whole place. The first round of employee placement to machines is done by using this method. In the second round, there are operatively solved cases of product moving to external hall, which causes "loops and bottlenecks of production". So, the second round defines re-alignment and search for the most proper and mutually compatible resources (employees, machines, material, tools and other axioms) for the most effective removal of those bottlenecks. All these operations are stored into a relational database, which defines all dependencies for the last steps in the production planning. The last steps means the handover of completed production plan, which fulfil all conditions for the most effective production and all available resources. These plans are handed over by the production hall chiefs to all supervisors, who forward them to their work teams and the production can start. Data concerning production process are stored in several databases on SQL servers. Data resources are:

1. Manual evidence of data by production dispatchers (employees, machines, parts of material balance).

2. Events, which are required for database administration (logged user, date of change, insertion date, update date, etc.).
3. Information systems outputs (TECHMNG, VOMNG, WIP – employee records).

3 Real Data Conversion and Pre-processing

In this section, we will focus on merging or conversion of data from several databases into one and on the data pre-processing for further process mining use. We will use data from one production process, namely transmission production.

These databases differentiate with the information system or company department, which is responsible for the storing of data. It means that, for instance, data concerning production orders, which originate from VOMNG information system are stored in a separate database excluding, e.g., technological procedures. To gain a dataset, which would consist of several activities from different data sources, we had to program a conversion software (extract from C# source code in Fig. 2). It took approximately 1 month to build the source code in a team of programmers. The software is a property of automotive company.

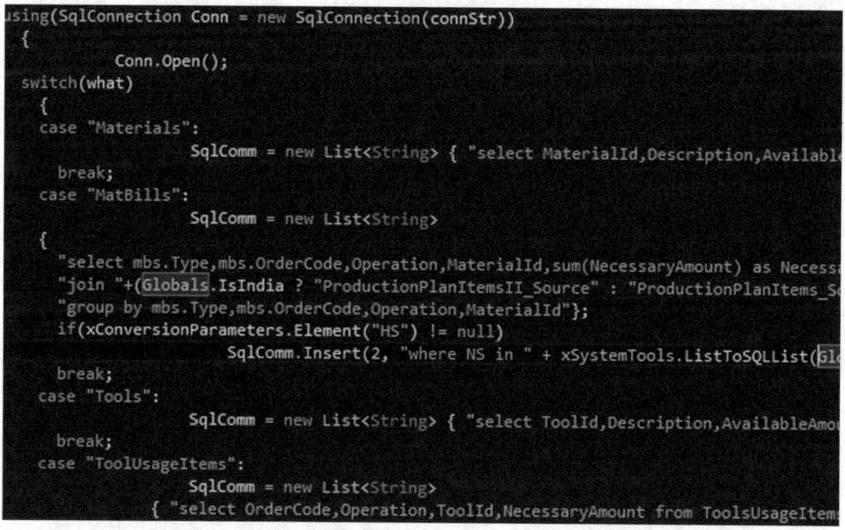

```
using(SqlConnection Conn = new SqlConnection(connStr))
 {
          Conn.Open();
 switch(what)
   {
   case "Materials":
                  SqlComm = new List<String> { "select MaterialId,Description,Availabl
     break;
   case "MatBills":
                  SqlComm = new List<String>
   {
     "select mbs.Type,mbs.OrderCode,Operation,MaterialId,sum(NecessaryAmount) as Necess
     "join "+(Globals.IsIndia ? "ProductionPlanItemsII_Source" : "ProductionPlanItems_S
     "group by mbs.Type,mbs.OrderCode,Operation,MaterialId"};
     if(xConversionParameters.Element("HS") != null)
                    SqlComm.Insert(2, "where NS in " + xSystemTools.ListToSQLList(Gl
     break;
   case "Tools":
                  SqlComm = new List<String> { "select ToolId,Description,AvailableAmo
     break;
   case "ToolUsageItems":
                  SqlComm = new List<String>
                { "select OrderCode,Operation,ToolId,NecessaryAmount from ToolsUsageItem
```

Fig. 2. Extract from conversion software. (Source: own)

Figure 3 illustrates the entire conversion process of real data to one dataset. Conversion of data begins with importing tables from three databases, which are used to store internal information about technological processes and machines (IS TECHMNG), employees and their activities (IS WIP), and finally about production orders and their timing (IS VOMNG). After that, we had to prepare desired tables and their structures manually to start automatic process of conversion. The convertor started

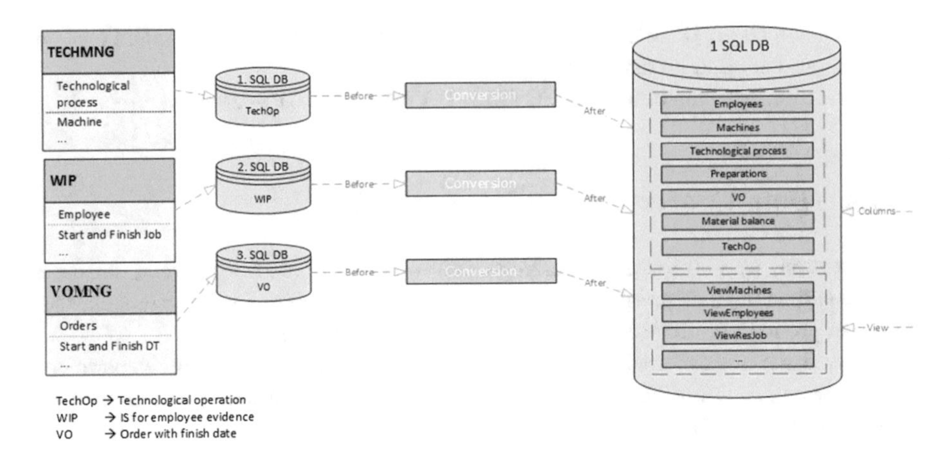

Fig. 3. The conversion process. (Source: own)

the collection of data from three databases and their tables and loaded required data into our pre-defined structures. The conversion process took some time, because we were loading XML records and additional information to employees and machines. After conversion, we have one SQL database stored with data, which belong to a production process in one economic centre. The next phase after conversion was the pre-processing of data.

For the majority of data evidence the automotive company uses information systems. This ensures minimal number of fault data entries. One of the most common faults or mistakes was, e.g., empty data entry of the name of the employee, which was responsible for a specific activity in a production process or a machine name. We corrected these entries with real IDs of employees or machines. The second most frequent mistake consist data entries with manually added values from employees. Specifically, it was problem with small and capital letters of same words, etc. In Table 1 several such mistakes are visible. For example, the machine with ID 6 is registered at location “ABC” with capital letters. On the other hand, the machine with ID 7 is located in the same place, but this can not be recognized, because of the data entry with small letters “abc”. Similarly, the machine with ID 10 record consists of same spaces in the filed location.

Table 1. Machines table – faults in some data entries. (Source: own)

M-ID	LOCATION	CODE	M-CATEGORY-ID
6	ABC	4	6
7	ABC	11	6
8	???	4	6
9	???	00C	7
10	A1	00D	???
…	…	…	…

Some of these common mistakes could be solved directly in conversion software using methods, which enable a change of letters (oLower, ToUpper) or erase blank spaces (Trim). Missing data entries should be filled manually. Tables 2, 3, 4 and 5 illustrate data after pre-processing.

Table 2. Data from IS TECHMNG. (Source: own)

VtchopId	Code	Description	Duration	Op-type	itemID
5483	340174/20	DRILL	3,3	PRODUCTION	1196
5518	340179/53	POINT QUOTE	24	POINT	1201
5523	340179/70	DRILL	9,24	PRODUCTION	1201
5544	340181/169	ADJUSTING OPERATIONS	0,75	SETTING	1203
5546	340181/179	ADJUSTING OPERATIONS	0,75	SETTING	1203
…	…	…	…	…	…

Table 3. Data from IS VOMNG. (Source: own)

PItemId	PID	PlanId	Count	Code	RequestedDeliveryDT
877	1209	2	100	340006	2015-10-19 00:00:00.000
878	1231	2	18	340007	2015-10-21 00:00:00.000
883	3044	2	50	340012	2015-10-14 00:00:00.000
889	5856	2	50	340018	2015-10-13 00:00:00.000
909	6353	2	200	340038	2015-10-19 00:00:00.000
…	…	…	…	…	…

Table 4. Machines data. (Source: own)

M-Id	Location	Code	M-Category-Id
6	ABC	4	6
7	ABC	11	6
8	ABC	65	6
9	ABC	00C	7
10	ABC	00D	8
…	…	…	…

Table 5. Employees data. (Source: own)

E-Id	Code	Name	Profession	EmployedFrom
1	1131	EMPLOYEE1	COMPTROLLER MECHANIC	1997-06-03
2	11815	EMPLOYEE2	GRINDER METAL	2009-12-01
3	1228	EMPLOYEE3	OPERATION AND ADJUSTMENTS	2015-05-04
4	13293	EMPLOYEE4	METAL TURNER	1983-01-01
5	13374	EMPLOYEE5	OPERATION AND ADJUSTMENTS	1983-12-24
…	…	…	…	…

4 Process Mining Analysis

Process mining analysis of our production process was performed in Disco by Fluxicon [7], which is a stand-alone tool for process mining analysis, with a focus on high performance (i.e., handling large and complex data sets) and ease of use.

The process mining algorithm of this tool is based on the fuzzy mining approach [8]. Disco supports seamless abstraction and generalization using the cartography metaphor. This way the tool is able to deal also with complex, Spaghetti-like processes. Other dimensions like performance can be analyzed through advanced visualizations of the mined process models [9].

After starting Disco and uploading our pre-processed log file in XLSX format, we have to assign columns from our source file to categories, recognized by Disco (case, activity, timestamp, resource, etc.). Here, we have listed the association of columns to categories (Fig. 4):

- Code – Case category
- Description – Activity category
- StartDT/FinishDT – Timestamp category
- EmployeeName - Resources category
- CategoryDescription, ProductDescription – Other category

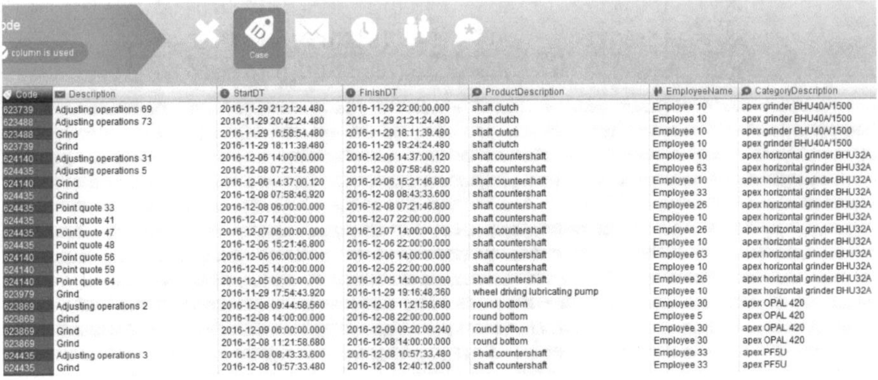

Code	Description	StartDT	FinishDT	ProductDescription	EmployeeName	CategoryDescription
623739	Adjusting operations 69	2016-11-29 21:21:24.480	2016-11-29 22:00:00.000	shaft clutch	Employee 10	apex grinder BHU40A/1500
623488	Adjusting operations 73	2016-11-29 20:42:24.480	2016-11-29 21:21:24.480	shaft clutch	Employee 10	apex grinder BHU40A/1500
623488	Grind	2016-11-29 16:58:54.480	2016-11-29 18:11:39.480	shaft clutch	Employee 10	apex grinder BHU40A/1500
623739	Grind	2016-11-29 18:11:39.480	2016-11-29 19:24:24.480	shaft clutch	Employee 10	apex grinder BHU40A/1500
624140	Adjusting operations 31	2016-12-06 14:00:00.000	2016-12-06 14:37:00.120	shaft countershaft	Employee 10	apex horizontal grinder BHU32A
624435	Adjusting operations 5	2016-12-08 07:21:46.800	2016-12-08 07:58:46.920	shaft countershaft	Employee 63	apex horizontal grinder BHU32A
624140	Grind	2016-12-06 14:37:00.120	2016-12-06 15:21:46.800	shaft countershaft	Employee 10	apex horizontal grinder BHU32A
624435	Grind	2016-12-08 07:58:46.920	2016-12-08 08:43:33.600	shaft countershaft	Employee 33	apex horizontal grinder BHU32A
624435	Point quote 33	2016-12-08 06:00:00.000	2016-12-08 07:21:46.800	shaft countershaft	Employee 26	apex horizontal grinder BHU32A
624435	Point quote 41	2016-12-07 14:00:00.000	2016-12-07 22:00:00.000	shaft countershaft	Employee 10	apex horizontal grinder BHU32A
624435	Point quote 47	2016-12-07 06:00:00.000	2016-12-07 14:00:00.000	shaft countershaft	Employee 26	apex horizontal grinder BHU32A
624435	Point quote 48	2016-12-06 15:21:46.800	2016-12-06 22:00:00.000	shaft countershaft	Employee 10	apex horizontal grinder BHU32A
624140	Point quote 56	2016-12-06 06:00:00.000	2016-12-06 14:00:00.000	shaft countershaft	Employee 63	apex horizontal grinder BHU32A
624140	Point quote 59	2016-12-05 14:00:00.000	2016-12-05 22:00:00.000	shaft countershaft	Employee 10	apex horizontal grinder BHU32A
624140	Point quote 64	2016-12-05 06:00:00.000	2016-12-05 14:00:00.000	shaft countershaft	Employee 26	apex horizontal grinder BHU32A
623979	Grind	2016-11-29 17:54:43.920	2016-11-29 19:16:48.360	wheel driving lubricating pump	Employee 10	apex horizontal grinder BHU32A
623869	Adjusting operations 2	2016-12-08 09:44:58.560	2016-12-08 11:21:58.680	round bottom	Employee 30	apex OPAL 420
623869	Grind	2016-12-08 14:00:00.000	2016-12-08 22:00:00.000	round bottom	Employee 5	apex OPAL 420
623869	Grind	2016-12-09 06:00:00.000	2016-12-09 09:20:09.240	round bottom	Employee 30	apex OPAL 420
623869	Grind	2016-12-08 11:21:58.680	2016-12-08 14:00:00.000	round bottom	Employee 30	apex OPAL 420
624435	Adjusting operations 3	2016-12-08 08:43:33.600	2016-12-08 10:57:33.480	shaft countershaft	Employee 33	apex PF5U
624435	Grind	2016-12-08 10:57:33.480	2016-12-08 12:40:12.000	shaft countershaft	Employee 33	apex PF5U

Fig. 4. Columns assigned to categories in DISCO. (Source: own)

After starting the import process a process map was created in the form as was expected (a part of the process map is depicted in Fig. 5). One can see logically connected sub-processes in the map. Every operation/activity and the entire flow of the product with frequencies in a graphic form is visible in the process. DISCO tool is also able to give us a functionality to investigate the production process from the statistics perspective. While the process map view gives an understanding about process flow, the statistics view provides an additional overview information and detailed performance metrics about the selected or filtered processes. The production process starts on November 29, 2016 and ends on December 14, 2016. The process consists of 490 events, 116 cases, and 248 activities with the mean time case duration 34.6 h. We get

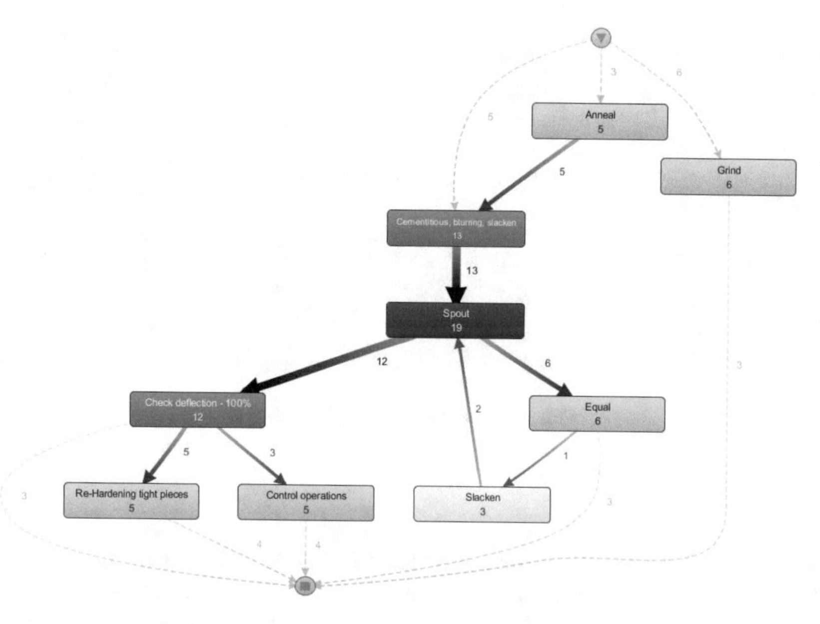

Fig. 5. Filtered process map. (Source: own)

to filtered process by filter function in the statistics tab in Disco. Different cases are visible in process map with different frequencies of activities and employees responsible for activities. A case is a particular process instance. We filtered a part of the process map, which depicts activity *Spout*, and several others connected to *Spout* (Fig. 5). *Spout* activity was performed 19 times with mean duration 58.4 min.

There are 3 activities preceding the activity *Spout*. Process map starts with *Anneal* activity, which was performed 5 times and 5 cases continued to *Cementious, blurring, slacken* (performed 13 times). There were also some other cases from different parts of the production process, which used *Cementious, blurring, slacken*. This activity was performed 13 times. The *Grind* activity was performed 6 times. Activity *Spout* was followed by 12 cases to activity *Check deflection – 100%* and based on its result, 5 products had to be repaired (*Re-Hardening light pieces*) and 3 products went to the last control operation (*Control operations*). 6 products from *Spout* followed to the *Equal* activity. These products had not to be checked for deflections. But some cases used the *Slacken* activity and followed to the *Spout* again. Check deflection activity mean time duration is 2.5 h.

A filtered process map in Fig. 5 with 19 cases, 113 events and 42 activities has a mean case duration of 54.9 h. However, the case duration statistics (Fig. 6) show that some of the cases (e.g., 624429, 624115, 623869, 355242) lasted more than 5 days. It indicates some possible problems with these cases. It could be a hint for production managers to concentrate on them among other options for process improvements. There can be as well analysed other aspects of described production process, e.g., activities, variants, events per case, case utilizations, resources, waiting time, etc. This will be a part of future research.

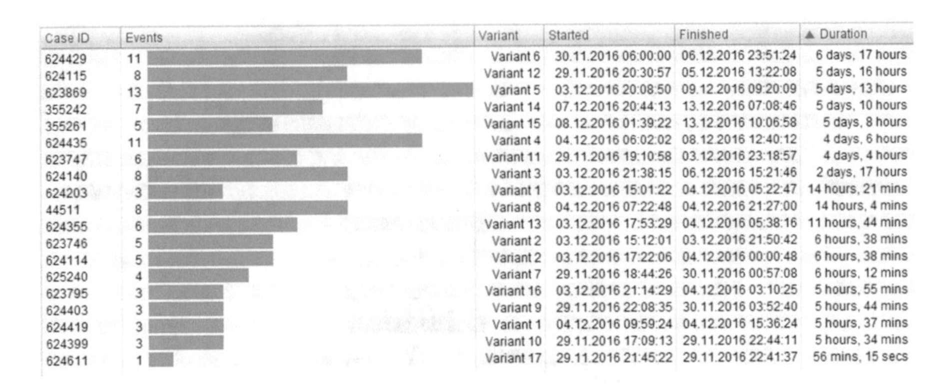

Case ID	Events	Variant	Started	Finished	▲ Duration
624429	11	Variant 6	30.11.2016 06:00:00	06.12.2016 23:51:24	6 days, 17 hours
624115	8	Variant 12	29.11.2016 20:30:57	05.12.2016 13:22:08	5 days, 16 hours
623869	13	Variant 5	03.12.2016 20:08:50	09.12.2016 09:20:09	5 days, 13 hours
355242	7	Variant 14	07.12.2016 20:44:13	13.12.2016 07:08:46	5 days, 10 hours
355261	5	Variant 15	08.12.2016 01:39:22	13.12.2016 10:06:58	5 days, 8 hours
624435	11	Variant 4	04.12.2016 06:02:02	08.12.2016 12:40:12	4 days, 6 hours
623747	6	Variant 11	29.11.2016 19:10:58	03.12.2016 23:18:57	4 days, 4 hours
624140	8	Variant 3	03.12.2016 21:38:15	06.12.2016 15:21:46	2 days, 17 hours
624203	3	Variant 1	03.12.2016 15:01:22	04.12.2016 05:22:47	14 hours, 21 mins
44511	8	Variant 8	04.12.2016 07:22:48	04.12.2016 21:27:00	14 hours, 4 mins
624355	6	Variant 13	03.12.2016 17:53:29	04.12.2016 05:38:16	11 hours, 44 mins
623746	5	Variant 2	03.12.2016 15:12:01	03.12.2016 21:50:42	6 hours, 38 mins
624114	5	Variant 2	03.12.2016 17:22:06	04.12.2016 00:00:48	6 hours, 38 mins
625240	4	Variant 7	29.11.2016 18:44:26	30.11.2016 00:57:08	6 hours, 12 mins
623795	3	Variant 16	03.12.2016 21:14:29	04.12.2016 03:10:25	5 hours, 55 mins
624403	3	Variant 9	29.11.2016 22:08:35	30.11.2016 03:52:40	5 hours, 44 mins
624419	3	Variant 1	04.12.2016 09:59:24	04.12.2016 15:36:24	5 hours, 37 mins
624399	3	Variant 10	29.11.2016 17:09:13	29.11.2016 22:44:11	5 hours, 34 mins
624611	1	Variant 17	29.11.2016 21:45:22	29.11.2016 22:41:37	56 mins, 15 secs

Fig. 6. Case durations. (Source: own)

5 Conclusion

The aim of this paper was to introduce the conversion of real data about production process from different information systems (log files) to the format, which is suitable for process mining analysis in a large automotive company. We introduced basic information about the entire manufacturing process and described the conversion and pre-processing of data to be imported into the DISCO tool for simple process mining analysis. Finally, we introduced some statistics results of the analysis to help us understand the control-flow and other parameters of the production process.

In the future, we will try to search for weak points in the production process and possibly, to export the process model to the XES format to use it in more sophisticated analysis in ProM tool.

Acknowledgement. This paper was supported by the project of Silesian University in Opava, Czech Republic SGS/19/2016 titled "Advanced mining methods and simulation techniques in business process domain."

References

1. Aalst, W.V.D.: Process Mining: Data Science in Action, 2nd edn. Springer, New York (2016). ISBN 978-3-662-49850-7
2. Gibert, K., Sanchez-Marre, M., Izquierdo, J.: A survey on pre-processing techniques: Relevant issues in the context of environmental data mining. AI Commun. **29**(6), 627–663 (2016)
3. Zakarija, I., Skopljanac-Macina, F., Blaskovic, B.: Discovering process model from incomplete log using process mining. In: Mustra, M., Tralic, D., Zovkocihlar, B. (eds.) Proceedings of 57th International Symposium ELMAR-2015, pp. 117–120. IEEE (2015)
4. Osborne, J.W.: Best Practices in Data Cleaning: A Complete Guide to Everything You Need to do Before and After Collecting Your Data. Sage, Los Angeles (2013)

5. Suriadi, S., Andrews, R., ter Hofstede, A.H.M., Wynn, M.T.: Event log imperfection patterns for process mining: Towards a systematic approach to cleaning event logs. Inf. Syst. **64**, 132–150 (2017)
6. Abedjan, Z., Chu, X., Deng, D., Fernandez, R.C., Ilyas, I.F., Ouzzani, M., Papotti, P., Stonebraker, M., Tang, N.: Detecting data errors: Where are we and what needs to be done? Proc. VLDB Endow. **9**(12), 993–1004 (2016)
7. Gunther, C.W., Rozinat, A.: Disco: Discover your processes. In: Proceeding of BPM Demos, CEUR Workshop Proceedings, Vol. 940, pp. 40–44 (2012)
8. Rozinat, A.: Disco User's Guide (2017). https://fluxicon.com/disco/files/Disco-User-Guide.pdf
9. Aalst, W.V.D.: Process Mining: Discovery, Conformance and Enhancement of Business Processes. Springer, New York (2011). ISBN 978-3-642-19344-6

Multi-Agent BPMN Decision Footprint

Towards Decision Collaboration Along Distributed BI Process

Riadh Ghlala[1,2(✉)], Zahra Kodia Aouina[1], and Lamjed Ben Said[1]

[1] Universit de Tunis, ISG, LR11ES03 SMART, 2000, Cit Bouchoucha Le Bardo, Tunis, Tunisia

[2] Higher Institute of Technological Studies of Rades (ISETR), Rades, Tunisia

riadh.ghlala@isetr.rnu.tn, {zahra.kodia,lamjed.bensaid}@isg.rnu.tn

Abstract. Nowadays, we are confronted with increasingly complex information systems. Modelling these kinds of systems will only be controlled through appropriate tools, techniques and models. Work of the Open Management Group (OMG) in this area have resulted in the development of Business Process Model and Notation (BPMN) and Decision Model and Notation (DMN). Currently, these two standards are a pillar of various business architecture Frameworks to support Business-IT alignment and minimize the gap between the managers' expectations and delivered technical solutions. Several research focus on the extension of these models especially BPMNDF which aims to harmonize decision-making throughout a single business process. The current challenge is to extend the BPMNDF in order to cover business process in a distributed and cooperative environment. In this paper, we propose the Multi-Agent BPMN Decision Footprint (MABPMNDF) which is a novel model based on both BPMNDF and MAS to support decision-making in distributed business process.

Keywords: BPMN · DMN · BI · MAS · MABPMNDF

1 Introduction

Business Intelligence (BI) is a business management term used to describe applications and technologies which are used to gather, provide access to, and analyse data and information about the organization, to help make better business decisions [1]. This activity of the company is materialized by business processes aiming to achieve the objectives established by the managers. Business Process Model and Notation (BPMN) which is a standard in its field, has been announced by the Object Management Group (OMG) in 2006 and is currently in version 2.0 BPMN since 2011 [2]. This latest version has introduced several new features to improve the modeling of business processes with a focus on several aspects such as:

G. Jezic et al. (eds.), *Agent and Multi-Agent Systems: Technology and Applications*, Smart Innovation, Systems and Technologies 74,
DOI 10.1007/978-3-319-59394-4_23

- Integration of new elements in the specification (events, gateways, activities, etc.).
- Distribution of business processes using diagrams of collaboration and choreography.
- Automating the business process execution by improving the BPEL XML based language (Business Process Execution Language).

Decision-making represents another field of investigation favoring the improvement of business process modeling. It was also an OMG center interest and has led to the invention of the Decision Model and Notation (DMN) in 2013 [3]. The valuable contribution of the DMN has encouraged the enterprise architecture community to cover other aspects of decision-making in business process. The challenge raised in several kinds of business process and especially in a BI process is to effectively manage decision-making in a case of distributed and communicating processes. In this context, the Multi-Agent System is considered as a suitable choice to model a solution allowing to satisfy this kind of need. This paper is structured as follows: Sect. 2 overviews the related work about decision-making and collaboration in business processes. Section 3 argues the choice of Multi-Agent Systems as decision-making modelling solutions in distributed business processes. Section 4 presents our approach that is the Multi-Agent BPMN Decision Footprint (MABPMNDF). Section 5 will be devoted to highlights the contribution of MABPMNDF in decision-making during BI process. Finally, we summarize the presented work and outline its extensions.

2 Decision-Making and Collaboration in Business Process: Related Works

2.1 BPMN and Decision-Making

A literature review on decision-making and its relationship to business processes showed a progression in dealing with decision. Indeed, the first preoccupation was concentrated on the separation between decision-making modelling and process modelling [4,5]. Research in this field are based on the collection, modelling and integration of business rules in business processes [6,7]. This work is crowned in the industrial world by DMN [3] which has become a standard in modelling decision-making in business process [8]. The second focus is the serialization of business rules and automation of its processing and its exchange [9,10]. Several open-source and proprietary software [11–14] appeared and are in competition to implement both of BPMN and DMN standards based on BPEL and FEEL languages [2,8]. According to OMG, business processes are modelled through two standards [8]: Business Process Model and Notation (BPMN) and Decision Model and Notation (DMN). The first is used to represent the various tasks and their relationships. The second supports decision making in the business process. Figure 1 shows an excerpt from a BI process in which we use these standards to model decision making at the data extraction task. The BPMN standard: it is the substrate for modelling business processes. It allows to represent graphically the company's activities to ensure better collaboration between managers and IT engineers [2]. The example given in Fig. 1 describes the ETL software selection task in a BI process. The decision-making is delegated to DMN [3] to work around the issue. The DMN standard: it is a BPMN add-in. It is structured in two distinct parts: (i) Decision Requirements Diagram containing the decision to study, business knowledge models, input data and knowledge source; (ii) Decision Logic which is represented by a decision table that can be converted into FEEL scripting language (Friendly Enough Expression Language) [8]. Thanks to this marriage between the BPMN and the DMN, decision-making in business processes has become an unquestionable acquis. This fact encouraged researchers to go further in the mastery of decision-making in business processes. Issues such as harmonization, uncertainty and collaboration of decision-making are currently top-level topics around decision-making in business processes. Figure 2 describes the BPMNDF [15], a model already proposed to ensure decision-making harmonization throughout the business process. Even if it is a collaboration between the phases of a single process and not a real distributed process or a set of communicating processes, BPMNDF [15] represents a first step towards collaboration in decision-making. It is a coupling of a BPMN with a novel DMN version based on additionally Decision Repository (DR) and Decision Memorization/Decision Regard (DM/DR) Algorithm responsible for managing the repository.

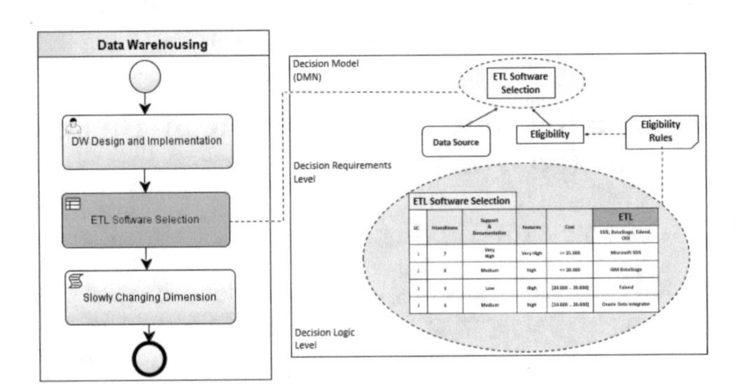

Fig. 1. Using BPMN and DMN in BI process

2.2 BPMN and Collaboration

The BPMN has provided support for collaboration since its preliminary releases. Indeed, the modeling of this need was simulated in BPMN 1.0 using swimlanes [2] whereas BPMN 2.0 introduced new diagrams according to the nature of this collaboration [2]. The new BPMN 2.0 diagrams are: orchestration diagram, collaboration diagram, choreography diagram and conversation diagram. We are interested in our research by the two diagrams: orchestration and collaboration. The first one is the BPMN Orchestration Diagram. This diagram depicting a sequence of coordinated activities from a single control center. It can be improved by

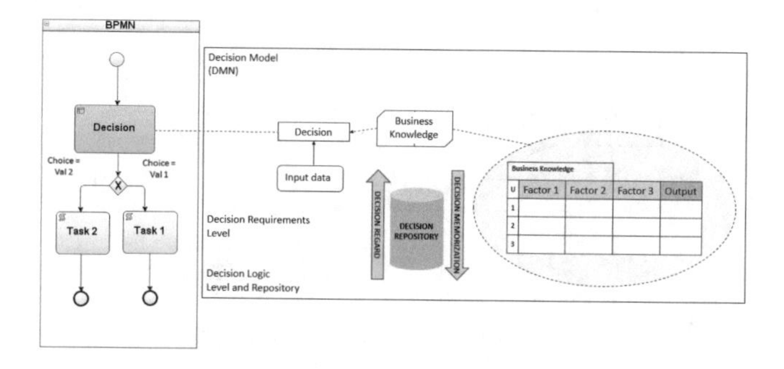

Fig. 2. BPMNDF [15]

introducing the swimlanes to highlight the organizational aspect of the process. In this case, the problem of the collaboration of decision-making in the business process does not arise because it is not a real distributed process. As shown in Fig. 1, the DMN perfectly supports decision-making at the task level and as shown in Fig. 2, BPMNDF [15], enhances decision-making through harmony along the process. The second one is the BPMN Collaboration Diagram. It is a diagram represents interactions between two or more processes, where each individual process reflects a person, role or a system. In this case, we need a new approach to manage decision-making in such a distributed and collaborative environment. An extension of the DMN is required to support not only decision-making at the task level, not also just the harmony of decisions throughout a single process but the challenge is to support decision making in multiple collaborative processes.

3 Decision-Making: From Support to Collaboration in Business Process

3.1 Decision Collaboration Necessity in Business Process

Collaboration is a natural need in business processes. The task of decision-making is, perhaps, the task most concerned by this collaboration. Because of, on the one hand the influence of the previous decisions on the current decision and on the other hand the impact of the current decision on the rest of the process. In each system, collaboration is governed by approaches that take into consideration organizational, technical, safety and performance aspects. In our research, the current challenge is how to combine the two diagrams: DMN and collaboration diagram to support decision making in a collaborative environment. In this paper, we propose an MABPMNDF model based on Multi-Agent Systems to meet this need.

3.2 Multi-Agent Systems as Modelling Solution

A Multi-Agent System is a set of software agents that interact to solve problems that are beyond the individual capacities or knowledge of each individual

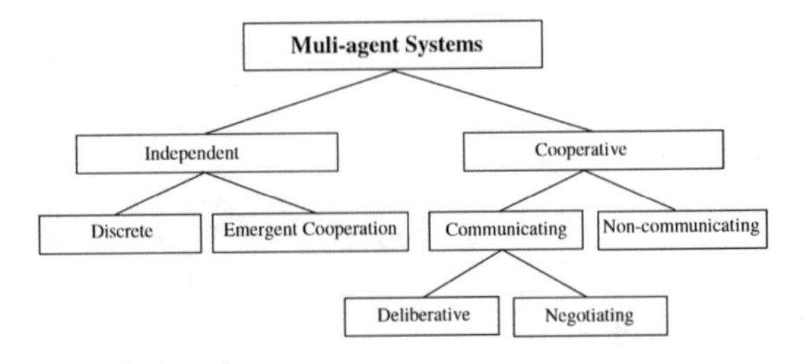

Fig. 3. Cooperation typology [17]

agent [16]. Objectives of this cooperation can be: achieving individual or common goals, labour division, task allocation, conflict avoidance, maximum reward, system integration, maintaining system functionality, system coordination, knowledge and information acquisition and/or sharing, collective intelligence. In our research, we are interested by the collaboration of decision-making. A goal perfectly ensured by Multi-Agent Systems. Several typologies of Multi-Agent Systems can be envisaged depending on the degree of synergy between the agents that contain it. Figure 3 shows a classification of these typologies based on criteria such as interdependence, communication and uncertainty. In this paper, we propose a novel Multi-Agent based model to support decision-making in business processes considered as collaborative, communicating and deliberative Multi-Agent Systems.

4 MABPMNDF Model

4.1 MABPMNDF Presentation

Decision-making in a business process is always based on business rules stored in a repository. The challenge raised by our proposed model is to ensure efficient management of this repository in a collaborative environment. Multi-Agent BPMN Decision Footprint MABPMNDF is a Multi-Agent Based System (MABS) which aims to give new functionalities to business process designers in order to support decision-making in a distributed and collaborative environment. It is an extension of the BPMNDF [15], a model that ensures only the harmony of decision-making between phases throughout a single business process. Figure 4 shows a sympathetic description of this model with its different agents around a fundamental component of the system which is the Business Rules repository.

4.2 Agents Description

The study of the agents of this model makes it possible to classify them according to two characteristics:

- The role of the agent which may be an administrative agent belonging to the management subsystem or a business agent belonging to the design subsystem.

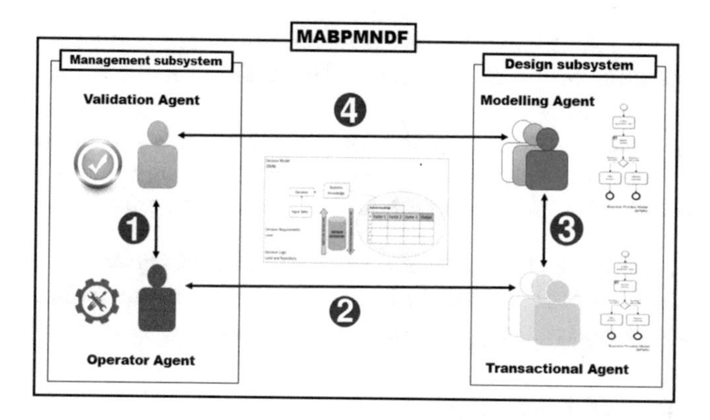

Fig. 4. MABPMNDF

- The degree of involvement of the business agent in the process through the privileges granted to it.

Management Subsystem Agents: The administration tasks of the model represent a crucial mission to ensure collaboration in a distributed business process. Thanks to agents in this management subsystem, we succeed in mastering the management of the repository, strengthening the reliability of business rules, controlling the security aspect and following versioning. We identify two types of agents in this subsystem:

- Validation Agent: It is the main agent of the system. It oversees the overall operation of the business-side system and delegates technical administration tasks such as system initialization and account management to the administrator agent.
- Operator Agent: It is a technical agent. It plays an intermediary role between the validation agent and business agents. The administrator agent is responsible, under cover of the validation agent, for the addition, modification and suppression of a business agents. It is responsible also for changing roles of these agents by the granting or denying their privileges. Finally, it enables and disables historical or contextual versions of the repository according to the needs.

Design Subsystem Agents: Decision-maker is called upon to accede to business rules stored in the repository in order to rationalize his decision. It can eventually add new rules to the repository or propose changes to existing rules. A member wishing interact with the deposit is considered as a business agent. All agents in the design subsystem are business agents. Depending on the degree of its involvement in the business process, the business agent can be a modelling agent or a transactional agent.

- Transactional Agent: is an agent whose only privilege is the read-only access to the repository in order to masters the decision-making task in business process.
- Modelling Agent: is an agent with more privileges than a transactional agent. It has business rule definition privileges such as adding, modifying or dropping. It collaborates with the transactional agents to enrich the deposit.

The overlap between the business agents and the administrative agents is possible although it is not recommended. Indeed, we can have members ensuring at the same time the business tasks and administration of the model.

4.3 MABPMNDF Message Protocol

Figure 5 shows an excerpt of the MABPMNDF message protocol via a UML sequencing diagram. It specifies the sequence of messages exchanged between the different agents of our model in order to request a new business rule.

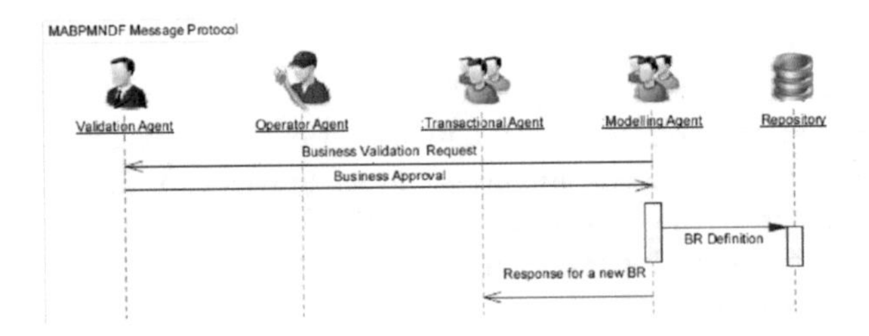

Fig. 5. MABPMNDF message protocol

4.4 Deployment Modes

In this section, we describe the different scenarios for using our model. The strategy is determined according to several parameters regardless of technical options like (i) infrastructure that can be desktop, web or mobile development (ii) business rules serialization that can be with XML, JSON, relational model or NoSQL databases (iii) user interface requirements that can be batch, interactive or near real-time. We insist in this first release of our model on the establishment of the repository. This parameter is relative to the creation and refresh of the repository. Two approaches are possible: an upstream creation or a downstream creation.

Upstream: The process starts from a pre-built repository and dispatch it to the different members of the team. Any changes will only take effect on the next release that will be released to the team. A rather cumbersome and less realistic approach for this is rarely used in information systems. The versioning concept in this approach reflects a repository update.

Downstream: The process starts from an empty repository and feed it gradually. The repository is built over time and it is transformed into a patrimony of the company. A realistic and very frequent approach in most information systems. The visioning concept according to this approach is seen as a contextualization between the type of business process and the repository used.

5 Simulation of a BI Process with MABPMNDF

5.1 Study Example

In this section, we provide an illustrative example of our MABPMNDF model applied to a BI process. The process encompasses the working of a start-up specialized in information systems engineering and specifically in the integration of business intelligence solutions. Knowing that a Corporate BI process is typically subdivided into three sub-processes: (1) data warehousing, (2) deductive and predictive analysis, and (3) reporting to develop reports, dashboards and scorecards. The studied company is then composed of three teams involved in this business process that can be described as a distributed process. In fact, each sub-process has its own geo-temporal parameters. Figure 6 illustrates the BI process using the BPMN collaboration diagram. The purpose of the simulation study is to understand how cooperation between the different members of the different teams is managed in terms of decision-making in the business process. This objective is materialized by the efficient management of the repository containing the business rules.

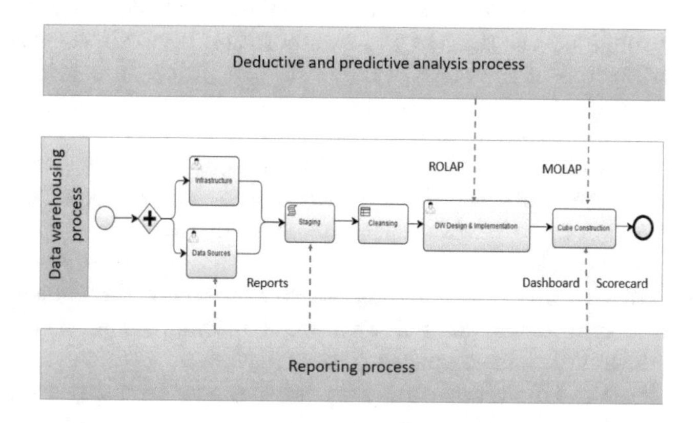

Fig. 6. BPMN collaboration diagram for BI process

5.2 Mapping Between MABPNDF and BI Process

The application of our MABPMNDF model on the chosen process requires in the first step an initialization phase triggered by a member of the company.

This member will be considered later as the validation agent who will oversee the entire process of his management and design side. This initialization is carried out by another member of the company. The last will be considered later as the operator agent. The mission of the operator agent is to implement the repository with the upstream or downstream strategy depending on the validation agent's setpoint. Thereafter it will be available to the other team members to act as the intermediary between them and the validation agent in order to respond to their requests for registration, profiling and activation of a particular version of repository. Each member of the team involved in the design of the BI process is considered as a business agent. It begins with being a transactional agent with read-only access on the repository. In the case of a need for a new BR or the modification of an existing BR. The transactional agent must collaborate with a modeling agent with such privileges to ensure its need. Let us note that a transactional agent can express its desire to become a modeling agent. If the request is accepted it will have additional privileges like the definition of business rules and not only their use.

5.3 Benefits of MABPMNDF on Decision-Making in Business Process

The assignment of our model is to facilitate and supervise at the same time the transactions around the repository of business rules for a rationalized decision-making in the BI business process. Indeed, a BI business process designer, whatever is his position in the process, can refer to reliable business rules to elaborate his decision. These decisions cannot be contradictory or unrealistic, because they are based on already validated rules. This takes into consideration the harmony of decisions throughout the process and collaboration between the various stakeholders. Contributions of the MABPMNDF in a BI process are several. We note especially:

- Avoiding inappropriate choices about technologies and tools used. Indeed, the repository can warn the designer through a business rule to streamline its ETL selection to be portable with its data sources.
- The focus on the principle of 'what we can do and not on what the business manager wants to do'. For example, we do not accept the choice of designing near-real time dashboards if we have a mechanism for extracting and refreshing data with widely spaced periodicity.
- Transforming the BI process into an adaptive way by using business rules derived from the real world with local parameters of the subsystem and not inappropriate business rules with generic parameters. Thereby, we must take into account the capacities and limitations of the infrastructure used in terms of memory and processing when implementing multidimensional structures or when choosing data mining algorithms.
- Strengthening the agility of the BI process by studying the feasibility of a suitable choice, even if the repository contraindicates it, with the modification of an existing business rule. The above-mentioned example concerning the

choice of a near-real time dashboard can be achieved by playing on the business rule controlling the extraction of the data from the sources to the data warehouse. Therefore, we win a highly-requested feature in business processes that is the agility of IT solutions to support the business layer needs in information systems. We hope that our model MABPMNDF contributes on the dream of business-IT alignment.
- Using versioning feature to manage BI processes in different contexts or with different modes: legacy mode and current mode. This feature is very handy when we design the same process with various levels of requirements or when we are called to attending inherited processes in the company.

These various requirements represent serious challenges for the success of BI projects. Thanks to MABPMNDF, these gaps may be increasingly controllable and BI projects could be more likely to achieve their objectives.

6 Conclusion and Future Works

In this article, we argue that decision-making in a business process is always based on business rules. A good perception and management of these business rules widely improve decision-making in business process. This task is further complicated in a distributed environment. Our contribution is in the context of the work of the OMG, in particular of their BPMN and DMN standards. We propose the MABPMNDF, which is an extension of BPMNDF, a model already proposed in the literature in order to improve decision-making by emphasizing the harmony of decisions throughout a single process. MABPMNDF is designed to support decision-making cooperatively in a distributed environment using the Multi-Agent paradigm. To illustrate the contribution of our model, a simulation of the MABPMNDF is applied on a BI process to show its benefits in decision-making in business process. Our future work will be scheduled on three axes. First, the implementation of a Framework to apply the MABPMNDF model. The second step is to validate this approach with case studies in the industrial environment. The third axis is the improvement of this approach by focusing on several aspects around the repository as more security through certificates, more availability by applying the mechanisms of backup and replication and ultimately more performance through optimization of business rules serialization.

References

1. Sperka, R.: Agent-based design of business intelligence system architecture. J. Appl. Econ. Sci. **VII**(3(21)), 326–333, Fall 2012. Spiru Haret University, Romania. ISSN 1843-6110
2. Object Management Group: Business Process Modeling Notation Specification 2.0 (2011). http://www.omg.org/spec/BPMN/2.0/PDF/
3. Object Management Group: Decision Model and Notation 1.0 (2015). http://www.omg.org/spec/DMN/1.0/PDF

4. Batoulis, K., Meyer, A., Bazhenova, E., Decker, G., Weske, M.: Extracting decision logic from process models. In: Zdravkovic, J., Kirikova, M., Johannesson, P. (eds.) CAiSE 2015. LNCS, vol. 9097, pp. 349–366. Springer, Heidelberg (2015)
5. Biard, T., LeMauff, A., Bigand, M., Bourey, J.P.: Separation of decision modeling from business process modeling using new Decision Model and Notation (DMN) for automating operational decision-making. In: Camarinha-Matos, L.M., Bnaben, F., Picard, W. (eds.) PRO-VE 2015. IFIP, vol. 463, pp. 489–496. Springer, Heidelberg (2015)
6. Kluza, K., Nalepa, G.J.: Towards rule-oriented business process model generation. In: Proceedings of the Federated Conference on Computer Science and Information Systems, FedCSIS, Krakw, Poland, 8 September 2013, pp. 939–946 (2013)
7. Bajwa, I.S., Lee, M.G., Bordbar, B.: SBVR business rules generation from natural language specification. In: Artificial Intelligence for Business Agility. AAAI Spring Symposium Series (SS-11-03), pp. 2–8 (2011)
8. Taylor, J., Fish, A., Vincent, P.: Emerging Standards in Decision Modeling - An Introduction to Decision Model & Notation in iBPMS Intelligent BPM Systems: Impact and Opportunity. Future Strategies Inc., Brampton (2013). ISBN 978-0-9849764-6-1
9. Benson, T., Grieve, G.: UML, BPMN, XML and JSON. In: Principles of Health Interoperability. Health Information Technology Standards, pp. 55–81. Springer (2016)
10. Ouyang, C., Dumas, M., ter Hofstede, A.H.M., van der Aalst, W.M.P.: From BPMN process models to BPEL Web services. In: Proceedings of the Fourth International Conference on Web Services, pp. 285–292 (2006)
11. Camunda workflow and business process management. https://camunda.org/
12. Signavio Decision Manager. http://www.signavio.com
13. Decision Management Solutions: DecisionsFirst Modeler. http://decisionsfirst.com
14. The Digital Enterprise Suite. http://www.trisotech.com
15. Ghlala, R., Kodia Aouina, Z., Ben Said, L.: BPMN decision footprint: towards decision harmony along BI process. In: Dregvaite, G., Damasevicius, R. (eds) Information and Software Technologies, ICIST 2016. CCIS, vol 639. Springer, Cham (2016)
16. Potiron, K., El Fallah Seghrouchni, A., Taillibert, P.: Multi-agent system properties. In: Fault Classification to Fault Tolerance for Multi-agent Systems. Part of the Series Springer Briefs in Computer Science, pp. 5–10. Springer (2013). ISBN 978-1-4471-5045-9
17. Glavic, M.: Agents and multi-agent systems: a short introduction for power engineers. Technical report, University of Liege Electrical Engineering and Computer Science Department (2006)

Author Index

G. Jezic et al. (eds.), *Agent and Multi-Agent Systems: Technology and Applications*, Smart Innovation, Systems and Technologies 74,
DOI 10.1007/978-3-319-59394-4

Z

Printed in the United States
By Bookmasters